Mindy Brubaker '77

The Mystery
of the
Great Swamp

DRAWINGS BY CARL KIDWELL

WEEKLY READER

CHILDREN'S BOOK CLUB

presents

MARJORIE A. ZAPF

The Mystery
of the
Great Swamp

ATHENEUM   NEW YORK

To my Mother, Grace A. Brown

The Mystery
of the
Great Swamp

1

THE LAST OF THE SUMMER NIGHT HUNG LIKE A
soft, dark cloud over the great Okefenokee swamp.
Though it was not yet light, Jeb was up early,
eagerly skimming over the swamp waters in his light,
little boat.

Again and again, he thrust his long, three-
pronged pole into the water and the little punt leaped
forward, then glided swiftly and noiselessly along.
In the pre-dawn, the boat seemed the only moving
object in the whole, vast swamp. On and on it darted,
cutting through the watery grasslands that stretched
out on all sides as far as the eye could see. On and
on, through a quiet, timeless world where only the

tinkle and splashing of silvery fish broke the absolute silence.

Even before the last star had faded from the sky, Jeb had penetrated far into the watery prairieland. Still he kept up his fast pace, for determination and strong excitement were driving him on.

Suddenly, though the narrow craft barely rippled the smooth surface of the water, it disturbed a lone heron that had been patiently fishing for its breakfast. Airborne, the large bird glided gracefully along on the early morning wind. A gentle wind that carried the pungent odor of the pines, the sweet smell of the Cherokee rose, the fresh fragrance of the tall grasses, and the musky scent of hundreds of swamp alligators.

Now as Jeb watched the first rosy hint of dawn streak silently across a pewter-gray morning sky, he knew, more than ever, how much he loved the great swamp. This strange and beautiful region of dark, mysterious cypress forests and open, sunny meadows. This half-drowned land where water lapped endlessly around the trunks of trees and forever flooded the wide grasslands.

Presently, Jeb realized that he was well on his way to the island. There was no need for such a fast pace. Relaxing, he paused for a moment to brush back a lock of sun-bleached hair from his forehead; then watched while the roaming wind stirred acre after acre of reeds and grasses into endless waves of graceful, rippling motion.

"We're going to make it today, Mac," he said softly. "Nothing can stop us now."

At the sound of Jeb's voice, there was a slight movement from the front of the boat, where a gangling, black puppy turned his head to gaze back at the boy. Besides his head, only the dog's tail moved as it thumped against the flat, weathered boards.

Jeb smiled down at the dog. Everything was going well. All the frustrations of waiting were gone, and he felt suddenly carefree and gay.

Gradually, Jeb turned the boat in an easterly direction. It was time to get his bearings. His eyes narrowed as he glanced over the vast marshland, scanning mile after mile of waving reeds and grasses.

Even in the half-light of early morning, he could distinguish, far in the distance, the dark forest he was seeking.

"There it is," he whispered to himself.

And there it stood. Stark and black against the pale sky, flanking the wide prairie on the east, completely dominating the flat landscape.

Deep in that forest, just a week before, he had discovered the island. Though he had seen it for only a few minutes, and then from a distance, he could remember vividly how it had glowed in the soft, evening light, like a green jewel in its setting of emerald green waters, a lake almost as lovely as the island. Without a doubt, it was the most beautiful part of the swamp that Jeb had ever seen.

Too late to explore it that night, Jeb had known he would come back as soon as he could. But that had meant a whole week of waiting. For his father had gone off on a week's trip into the swamp with an archaeologist to look for Indian relics. With all the family's chores on his shoulders, Jeb had had no time for a trip. Not until today.

With the forest in sight, Jeb corrected his bearings once more, swinging his boat around until it pointed almost directly toward the mid-section of the dark forest. Determinedly, he sent the small punt onward, over the tips of the saw grass rooted in the black, mucky bottom; then in and out of narrow lanes blazed by countless alligators.

For a few moments, Jeb was completely lost in his thoughts, wondering again at the unusual fascination the island held for him. Was it entirely the great beauty of the spot? Or was it because he felt sure he was the first person to see the island and look upon its beauty? Whatever the reason, it was strong and compelling and it was luring him back.

Gradually, without realizing it, Jeb fell into a strange mood. As he gazed at the distant forest, something of its brooding and forbidding presence reached out across the waters and touched him. For endless moments, the whole swamp seemed to hang hypnotized in a dark spell.

Then the sun, like a fiery, red ball, rose slowly over the rim of tall trees edging the forest. Vivid color burned and flamed across the eastern sky. And

the strange mood was quickly broken.

With the rising of the sun, the swamp quickened miraculously into life. Its stillness was broken by the splashing of countless waterfowl and the singing of many birds. Unseen insects buzzed incessantly. Small, green katydids balanced delicately on the tips of slender grass leaves. And plump frogs leaped in and out of the water, exchanging one thick lily pad for another.

Suddenly, a large, brown object caught Jeb's eye. "A deer! And what a beauty!" whispered Jeb. "Don't make a sound!"

The dog remained silent while the boat glided, undetected, past a large buck feeding on succulent roots beneath the water. Jeb was delighted at seeing the animal at such close range, but, above all, he was pleased that Mac obeyed so well.

A dark shadow moved in the water. A large alligator slid silently across the channel in front of the boat, then submerged quickly out of sight.

A little prickle of worry crossed Jeb's mind. Soon the warmth of the sun would draw countless alligators from the depths of their small pools. They would swim under and around the little boat. They would crawl, by the score, onto the mud banks to bask in the sun. Jeb knew they would never attack him while he stayed in the boat. After all, he had been navigating among them all his life. But it was different today. Mac was with him.

"I know I've trained him well enough," Jeb told himself. "I'm sure he won't panic."

But deep in his heart, Jeb was not sure. He hated to admit it, but he really did not yet trust his dog.

He tried to push out of his mind all the stories he had heard of boating accidents, where a dog, or a boy, or a man had gone overboard in the path of an alligator. In his mind, he kept hearing the powerful jaws crunching down on its victim. He found himself watching Mac very closely, felt himself actually willing the dog to remember his training and obey.

Entirely unaware of Jeb's tenseness, the lanky ten-month-old black puppy sat crookedly on his haunches in the front of the boat. His silken ears were flattened along the sides of his sleek head so he would miss none of the exciting sounds of the swamp, and his moist nose twitched eagerly as he breathed in its many odors.

His steadiness now, his willingness to always sit in the same spot in the bow of the boat, was the result of months of training, begun almost the first day he had been Jeb's.

"Teach him from the start never to jump or move about," Jeb's father had warned then. "As small as he is, he can capsize your boat. If that happens, you'll be in real trouble."

And Jeb understood that the dog had to be taught always to mind at a moment's notice, for if he disobeyed or panicked, it could mean injury or death to both of them.

The small puppy had needed short lessons at first, lessons that had to be repeated over and over again for weeks at a time. It took all those weeks for Mac even to begin to lose his instinctive fear and distrust of snakes and alligators.

Every day, boy and dog had gone out in the swamp together. At first for only a few minutes a day, then an hour, then more. Each day, Jeb had watched Mac's eager intelligence and quick obedience sharpen and grow.

Still, the gnawing fear remained that Mac might not yet be ready for the long, hazardous journey to the island.

As Jeb thought about the coming trip along the dark, narrow run in the forest, he worried about the deadly snakes that twined themselves around bushes and branches and twigs. Mac still hunched a little to one side or the other whenever the boat occasionally passed close to a cottonmouth. On the trip to the island, snakes would be hanging from tree branches on both sides of the run. How would their continued presence affect Mac? Would he become so frightened and confused he would upset the boat? Would he try to jump overboard into the path of an alligator waiting to snap up anything within reach in its powerful jaws?

There was no getting around it; the trip to the island would be a severe test of Mac's obedience and training.

Despite his misgivings, however, Jeb had no intention of turning back. Gradually, the seemingly endless reeds and grasses gave way to acres of brown cattails that stretched clear to the great stand of cypress. Then, at last, looming just ahead, was the long, black mass of the forest.

For one brief second, Jeb glanced back at the miles of flower-decked meadowland he had left behind, and, for a moment, felt a twinge of regret at leaving those open, sunny spaces. Then, he brought his attention back to the dark, shadowy, half-drowned wilderness just ahead.

As he moved his boat slowly along the edge of the forest, he watched for guideposts he had learned to know by heart. There was the channel entrance to the fishing spot; next was the old log, wedged between a group of cypress knees; then past the hummock where the large magnolia grew. And always, he scrutinized each and every tree, looking for familiar landmarks that would pin-point his position. It was second nature for Jeb to watch the trees. Only by careful observation could he find his way from one spot to another and then safely back home. He had learned this skill gradually from the time he was a small boy. And now, vast areas of the swamp were home to him.

He searched eagerly for the landmark that would show him where the narrow boat run started, leading through the cypress forest.

"I must be getting close," he muttered to himself. "I remember that it's just past that small island."

Then, he saw what he was looking for. In front of him stood a tremendous tree. A tall, dead giant of a tree. Long ago it had been struck by lightning. Now all that remained was its gaunt, weathered trunk and gray, ghost-like branches completely shrouded with long streamers of Spanish moss.

As Jeb approached the old, dead sentinel, he threw back his head, straining to see the massive nest he knew was woven into its topmost branches. There, high up in a remote eyrie, a pair of great bald eagles returned year after year to nest.

Suddenly, an angry scream shattered the silence, and Jeb saw one of the large birds circle round and round, before disappearing into its shadowy hideaway.

Just a week before, late in the afternoon, Jeb had been drifting idly along at this same place, peering as now toward the top of the dead cypress to see if he could get a glimpse of the eagles on their nest.

But a loud, crashing noise had drawn his eyes to the center section of the tree, where a tremendous, black bear was thrashing its way downward after raiding a beehive. Clumsily, but quickly, it clawed its way to the ground. At that moment, Jeb's boat came so close to the bear, he could see its black muzzle, sticky with stolen honey.

He could also see the dark, angry cloud of bees buzzing around the bear's head.

Then things had happened fast. The bear, anxious to shake off the vicious swarm of stinging bees and startled by Jeb, had plunged headlong through the thick tangle of vines and bushes at the base of the old tree. A moment later, Jeb's boat had come abreast of the gaping hole the bear had torn in the green wall. Peering into the dimly-lit region beyond the opening, he had made a startling discovery. Through the tremendous hole, a green-tinged boat run was visible, that wound its way far back into the shadowy cypress forest. Jeb had been by the entrance to the run many, many times before, poling past it during every season of the year. Often, after leaving the fishing spot in the forest, he came south a mile or so this way to study birds. Yet, he had not known such a run existed, for it had been hidden by years and years of primitive, undisturbed growth. In fact, from the size of the torn vines, some as thick around as a man's wrist, he was positive the channel had never been used before.

Jeb had the sobering feeling that, years ago, time itself had sealed off the run with those thick vines and lush plants. Sealed it off and then forgotten it.

Now, with the green curtain of jungle growth torn away, the run looked like any other narrow channel that moved into the flooded cypress trees. And yet it was not like any other run. Somehow even in that first glance, Jeb had sensed this. In the first place, it moved more directly, at least at first, into the forest. It did not make so many twists and turns as most channels did, except maybe the one to the fishing spot. It was fairly straight. But the difference was even more than this. There was a sense of aloneness about it, an eerie quiet that set it apart.

2

JEB PAUSED AT THE ENTRANCE TO THE RUN. ALthough he was impatient to begin the long journey through the forest, he kept his boat motionless by anchoring the boat pole firmly in the mucky bottom. After the bright sunlight of the open prairieland, he needed time for his eyes to adjust to the dimly-lighted region of the cypress bay.

Gradually, his eyes became accustomed to the gloomy light of the forest. Slowly, he eased his boat down the narrow run. Though he moved cautiously, the dark trail ahead held no terror for him. He had been roaming in and out of such shadowy, greentinged cypress bays since he was five years old. He knew that this run led to a wild, remote, unexplored region, but he had traveled it once, and he was confident he would not lose his way.

The little boat slid farther and farther into the forest, silently gliding over the amber-colored water of the run as it went winding in and out among the twisted cypress trees.

In the thin, greenish light, the cypress trees loomed above Jeb, crowding and pressing in on him from all sides. Everywhere there were thick, endless groves of close-growing cypress, standing with their roots deep in the ancient swampland, and their swollen knees protruding above the dark water. They were tall trees that reached far up, stretching toward the light, where their thick, spreading tops massed together to shut out most of the sun's rays.

On both sides of the narrow run, thorny thickets of huckleberries and Cherokee rose formed an impenetrable wall of green. And all about, on every limb, every branch, every twig of the stately cypress, filmy streamers of silvery-gray Spanish moss waved hauntingly to and fro.

Jeb could see every detail now, and, just as he remembered, the cottonmouth snakes were everywhere. They were twisted and draped over the bushes that lined the run. They were coiled together in masses on the knobby knees of the cypress. They were hanging on low limbs and branches. As the boat glided along, the snakes closest to the run slid into the water and disappeared. Still, their very presence disturbed Mac. Jeb could sense his deep fear. He could see the tenseness in the dog's hunched shoulders.

Jeb spoke soothingly to help him control his fear.

"Steady, boy. Steady. Everything's all right. When those snakes hear us coming, they'll get out of the way."

Still, for all his show of calm, Jeb shuddered a little, too.

The small boat glided on and on, winding soundlessly down the slender thread of current, deeper and deeper into the forest. Here the air was cool and moist, and pungent with the smell of dense, green masses of maiden cane that were clumped along the sides of the run. The subdued green light, that filtered through the green roof of the arching trees, cast a cathedral-like spell over the bay.

But even though the light was thin and diffused, the smooth, dark water was like a tremendous mirror. Jeb could see his own clear reflection as he lifted his long pole in and out of the water. He could see perfect reflections of bushes, plants, and even the tops of the towering trees festooned with ghostly strands of moss. And, as he went gliding silently along, he had the eerie sensation that he was riding upside down.

But only minutes later the reflections ceased, for the trees became so thick that almost all light was trapped in the overhanging leaves. Before it had been twilight; now it was as black as night.

Jeb wished he could pay more attention to Mac, but the channel narrowed at this point and he had to give all his attention to threading his boat through the tangled cypress roots. He knew he was in the darkest and trickiest stretch of the run, but he remembered that in just a little while the run widened again.

Then suddenly, almost magically, a broad beam of sunlight pierced through the thick, matted tree tops. It penetrated down to the run itself, where it sparkled and danced on the dark water.

Surprised at the sudden brilliance, Jeb traced the golden ray up from the water to where it shimmered on a smooth, wet cypress knee. Then he followed it to a low limb that hung over the water a short distance away. But his eyes never got any farther.

His glance froze on that branch and fear flooded through him. For the golden beam spotlighted the body of a huge brown snake coiled grotesquely around the branch that jutted out directly over the boat lane.

In a few seconds, the boat would pass right under the monstrous, menacing snake.

Jeb was already so close he could see the chunky head drawn back ready to strike. He could see the snake's wide-open mouth with long fangs and gleaming, cotton-white interior. He could see its glaring, sinister eyes.

Jeb knew for sure that it would strike out viciously at anything it could reach. And Mac saw the snake, too. His body tensed. Every hair on his back bristled with fear. Every instinct urged him to jump clear of the boat. But now, because of weeks of careful training, he looked back at Jeb for a command.

Jeb didn't have much time. Quickly, he held up one hand.

"Sit!"

Then he prayed silently that Mac would obey.

In the face of such danger, Jeb acted intuitively. Bracing his feet against the sides of the punt, he jabbed his long pole into the water. He had to stop the boat. He couldn't let the cottonmouth reach them with its deadly fangs.

Jeb's arms ached with the sudden impact of wood against swamp bottom. Then fear became a hard knot pressing against his ribs. For, while he had managed to slow the forward motion of the boat, he had been traveling too fast to stop it completely. The little punt still moved inevitably forward.

Jeb's heart thumped madly against his chest. He had to frighten the snake out of the tree, for he was certain he could not stop the boat in time.

Gripping the boat pole with both hands, he frantically chopped the surface of the water—up and down, up and down.

The noise exploded on the water and then went echoing through the silent forest. Jeb was certain it would frighten the snake from the tree, but the cottonmouth still clung to the overhanging limb. Sweat broke out on Jeb's forehead. He struck the

water again and again. The heavy pole slipped in his sweaty hands and he almost lost his grip on it, but he managed to regain his hold and chopped the water again. And all the time, the question burned away in the back of his mind—would Mac obey his command?

Then, just when the situation seemed hopeless, the snake dropped into the dark water with a loud splash and slithered away.

Jeb let his breath out slowly in one long sigh of relief. It had been a close call, for the snake had landed in the water just a few inches in front of his punt. A half second later and it would have fallen into the boat where it could have reached either Mac or him.

Now, even though the danger was over, Jeb was cold and trembling, and he wondered if his legs would ever stop shaking. Even his arms felt heavy and weak; so he laid the pole in the boat and let it drift slowly along on its own.

One thing was sure, he reflected with relief. He need never worry about Mac again. Though the dog had been almost underneath the ugly, threatening snake, he had remained in his place in the boat. He had obeyed the command.

Jeb talked on and on to Mac to ease the tension that gripped them both.

"You're the best dog in the world, Mac, for minding so well. I told you those cottonmouths would get out of the way when they heard us coming. Why, if you had jumped in the water, a gator might have gobbled you up."

As the run widened, dim, greenish light once more filtered through the treetops, and Jeb could see the strange, enchanted world of the cypress forest again. He picked up his pole and pushed on.

Occasionally, long, filmy wisps of silvery moss brushed softly against Jeb's face, but he scarcely noticed it, there was so much to see. Now and then, brown logs appeared briefly in the channel; then changed unexpectedly into alligators with lowered lids, who slipped quietly away from the boat as it approached.

Several times in the dim, shadowy light, Jeb saw dark shapes moving silently among the gnome-like cypress, but what they were, he could only guess. For before Jeb could see them clearly, they plunged secretively into a background of even darker shadows and lost themselves in the unexplored depths of the jungle.

Presently, as Jeb rounded a bend in the run, he suddenly realized how hot and thirsty he was. So he stopped at the base of a large cypress tree for a drink. Stooping low in his flat-bottomed boat, he dipped a battered tin cup beneath the smooth surface and brought up a cupful of clear water that was stained a golden brown from some ingredient in the cypress trees. But he almost spilled the water from his cup when a shy bittern sprang from its hidden nest in the dense maiden cane beside the run, and rose into the air with a great flutter of wings and a sad, anguished cry. Then, barely a second later, a brilliant woodpecker with black and white wings and a scarlet crest swooped low over his boat, before melting into the shadowy depths of the jungle, where its sharp, staccato hammering punctuated the long silence of the forest.

Suddenly, the boat lurched sharply to the left, and Jeb felt it being pulled by the water into a side channel that opened up through thick berry bushes. Jeb clenched his hands around the pole, immediately alert to the dangers of the strange, mysterious currents that sometimes appear unexpectedly in the Okefenokee. He quickly guided his boat back onto the known waterway. He was no coward, but he knew what could happen if he lost the main trail, for such untried channels often led into wild regions of the swamp where men became hopelessly lost.

Because of the danger of getting lost in the Okefenokee, many, many square miles in its rugged, inaccessible heartland had never been explored. In fact, almost two-thirds of its 412,000 acres had not even been surveyed. Except for a venturesome few among them, even the swamp men went only to areas around its edges that they had learned to know from childhood.

All his life, Jeb had heard stories of people losing their way in the swamp. The lucky ones were found by search parties, or managed after days or weeks of an exhausting, terrifying ordeal to grope their way out. But Jeb knew there had been many unlucky ones who never returned, whose bones would lie forever in some primitive region of the great swamp.

Safely back on the right channel again, Jeb breathed a sigh of relief and continued on his way. He soon reached a sharp bend in the run, and knew that the lake and the island were not far away. From that point onward, the channel widened and the cypress trees became less dense. Bright sunshine seeped through the overhanging leaves and dappled the dark waters with great splotches of golden light.

Then, for one brief second, he even glimpsed the sparkle of the lake itself through the thinning cypress. And, though it was only a momentary, quick, bright gleam, it set Jeb's heart pounding even more rapidly against his ribs.

"We're almost there!" shouted Jeb.

Though his voice rang out loud and clear in the still air, its sound was almost completely drowned by the hoarse, insistent cries of a flock of great Canada geese that flew unseen above him, racing him to the lake over the thick, moss-covered tops of the cypress trees.

Their wild cries only added to Jeb's excitement, and he took his eyes off the run while he peered intently through the trees in the hope that he would catch another glimpse of the lake.

Thus, he didn't see his boat slice across a long spray of red roses that spilled out across the water in a wide streak of flaming color. Color that waved gently on the water like a vivid, scarlet flag of warn -ing. Instead, the punt cut the brilliant red gash in two. And, because of Jeb's preoccupation, the warning went unheeded. Moments afterwards, the sweet, cloying fragrance of crushed blossoms scented the moist, heavy air, but it wasn't until his boat slid over a cypress root and into a shallow mire, where it stuck fast in the thick, black mud, that Jeb finally remembered to bring his eyes and attention back to the channel.

Then, in desperation, he pushed fiercely with his pole against the soft mud. Pushed and strained with all his strength until, at last, the little boat came free; skimmed lightly over the last of the run; then whirled and burst in a great shower of white foam, out onto the sparkling waters of the open lake.

3

JEB GAZED SPELLBOUND AT THE EMERALD WATER
that sparkled and glistened before him. In his delight, he let his boat bob up and down on the bright
water, and squinted across its shining surface to the
tree-covered island that nestled at the far end of the
lake. For a long time, he feasted his eyes on the
fresh, green beauty; the primitive, untouched loveli-
ness of the island that had lured him back. Then
slowly, he surveyed the rest of the region.

The lake itself was a perfect oval, completely
encircled by a thick forest of dark cypress, dense
bushes, and lush ferns, all crowding close to the
water's edge. All about were hues and tones of

25

blending greens, from the silvery, gray-green of the Spanish moss, to the olive-greens and blue-greens of the bushes, to the deep, emerald-green of the tall ferns. At the water's edge, the ferns cast their dark emerald reflections upon the water, stamping it with the same deep emerald color; making the lake a rich, green liquid with hidden depths and lights that sparkled like a precious gemstone. As Jeb watched the shining color of the water, it made him think of a beautiful jewel, and, almost without realizing it, he gave the lake its name.

"Emerald Lake," he said softly to himself. "I'll call it Emerald Lake."

Gradually, however, Jeb became aware of a strangeness about the lovely lake cradled in the forest. Something he could not quite put his finger on was oddly disquieting. What was it? For a while the answer eluded him, but finally he realized what it was. Incredible as it seemed, he could see no exit or entrance to the lake except the one through which he had come. Puzzled, he scanned the thick forest surrounding the lake, again and again. But there were no visible runs leading from it. Other lakes in the swamp narrowed out into numerous waterways that led, in labyrinth fashion, from one dim cypress bay to another. Not all of them were clearly visible from any one spot, but generally at least two or three could be seen. Emerald Lake, however, appeared completely cut off from all other parts of the swamp.

As Jeb realized the complete isolation of the lake and the island, a wave of loneliness swept over him. Intuitively he glanced down at his dog. But his mood of loneliness was soon interrupted when a little wood duck scooted across the water in front of him, screeching alarm at the sight of his boat. The alarm startled a whole multitude of ducks—buffle-heads, pintails, mallards, canvasbacks, and mergansers. Then, while Jeb watched in wonder, hundreds of birds flew up from the water and, with a great flapping of wings, soared toward the other end of the lake.

Jeb's longing to explore the island was stronger now than ever before, and he groaned inwardly as he suddenly remembered one last chore he had to do before going to the island. He had promised his mother the night before that he would get enough fish for the evening meal; and, anxious as he was to explore the island, a quick glance at the rapidly rising sun decided him on his first course of action. Knowing that early morning was the best time for fishing, he resignedly pushed his punt toward the middle of the lake.

Floating free, Jeb cast out his fishing line time after time. Occasionally, the smell of pines drifted across the lake from the island, causing him to glance wistfully in that direction. But after a bit, he became so absorbed in his fishing that he scarcely noticed his surroundings. While Jeb pulled in one large fish after another, Mac curled himself into a round, furry ball and dozed peacefully in the bow of the boat.

During the time Jeb fished, his boat slowly drifted closer and closer to the island. When he finally looked up, after landing the largest fish of the morning, he was startled to see that he was a scant fifty feet from its shoreline.

Jeb could stand it no longer. Hurriedly, he scooped up several cupfuls of lake water, poured it over the fish in his bucket, and then eagerly sent his punt skimming toward the island.

Like most of the islands scattered throughout the great swamp, it was covered with a matted growth of swamp oak, cypress, and young pines, while crowding between the trunks of the trees was a riot of huckleberry bushes, gallberries, and sedges.

Slowing down, Jeb peered into the shadowy depths along the shoreline, wondering what wild animals might be peeking out at him from their secret hiding places among the brush, for he knew that bear, deer, and wild boar abounded in the forests of the larger islands. But the only sign of life he saw was a small, gray squirrel, leaping among the lacy leaves of an ancient oak, and a little, striped garter snake sunning itself on a lily pad.

Jeb guided his punt nearer to the island, edging his way so close to the shoreline that the boat was scraped on both sides by the flat, sword-like leaves of the wild, blue iris that grew there in profusion. Nearer and nearer he went, carefully searching the shoreline for a place to land, while his boat slid past ferns, sundews, and tall, spotted pitcher plants that were quietly ensnaring insects in their tube-like leaves. Closer and closer he inched, until he could see the pale blossoms of the tupelo trees.

Now and then, bird songs sweetened the unearthly quiet of the place, while occasionally, quick wings flickered in and out of deep shadows cast by the trees standing thick in the forest.

Suddenly, Jeb jabbed his pole into the water to stop the boat, for his sharp eyes noticed a huge mound of earth a short distance back from the shoreline. It was a mound about ten feet high and four feet wide, weathered smooth by countless seasons that had come and gone in the swampland.

With a shock, Jeb realized that he had found the ancient burial place of one of the early Indian inhabitants of the swamp. Though many similar mounds had been discovered on larger islands throughout the Okefenokee, this was the first time Jeb had seen such a burial site in his part of the swamp.

Now, gazing at the huge mound, while the water lapped softly around his boat, he reflected on the odd coincidence of his finding such a burial site at a time when his father and the archaeologist had been excavating just such a mound in another part of the swamp.

He thought, too, of the tales that had been so much a part of his early boyhood, tales that his grandfather had told about the early days of the Okefenokee, when it had been a great hunting ground for Indians.

Jeb knew that for generations legends had been passed down about a mysterious tribe of tall Indians who were thought to have been the first inhabitants of the great swamp. According to the legends, they had been a handsome race of near-giants, who were expert hunters, fishermen, and craftsmen, and who supposedly had lived on an island in a beautiful lake.

It was this handsome, intelligent race of "tall ones" who were credited with building the burial mounds for, in one site that had been excavated, the skeleton of a man over seven feet tall had been found.

But so many, many years had passed since the mound-builders lived in the swamp that no one really knew much about them. There were countless questions still to be answered: Where had they come from in the first place? And why, and to where, had they disappeared? In all the stories that Jeb had

heard, it was taken for granted that invading, war-like tribes of Indians, and later the white colonists who settled around the perimeter of the Okefenokee, had caused the tall Indians to retreat to their island fastness, and then, finally, to leave the swampland forever. But whatever the answers to the questions, Jeb knew that now only the huge burial mounds remained as tantalizing evidence of the mysterious tribe's life in the days gone by.

The sight of the mound overpowered Jeb with a renewed awareness of the antiquity of the great swamp, and he longed to examine the site more closely. Also, in some curious manner, the long shadows of the past cast a timeless spell over the lake, creating within him a feeling of strong kinship to the vanished race that had once roamed the meadows, cypress bays, and lakes just as he did now. But search as he would, Jeb found no suitable place to moor his boat, for the bank rose steeply from the water in one long, unbroken line.

Slowly Jeb continued his journey, swinging along the shoreline, looking for a small cove in a sheltered bend where he could beach his boat. But his punt rounded one gentle curve after another without success.

Drifting noiselessly along, Jeb became more and more aware of the immense, brooding silence that hung over Emerald Lake. An intense and eerie silence, it heightened his feeling of uneasiness and, unexplainedly, honed all his senses to a razor sharpness.

Slowly, his eyes moved along the gentle curve of the shoreline, but all he saw were the silent shadows of the forest as they drifted out to meet the lake. Realizing that he was scarcely daring to breathe, he forced himself to take a deep breath.

The silence made each sound that arose from the water or surrounding forest stand out sharply and clearly. There was no blending of sounds.

Somewhere a solitary limpkin cried its lonely call. Once a dark shadow passed slowly over the water and, as Jeb squinted against the sun, he saw an osprey hang momentarily in the sky before diving with dizzy speed into the water nearby. When the fish hawk broke the water, the sound was shattering, and even its wingbeats sounded as it flew to a tall tree with its catch.

Jeb found himself sifting and sorting every sound into its proper category. He made a mental note of each frog that croaked, each bird that winged its way up from the water. Even the small, unseen creatures that occasionally splashed into the water plinked their way into his conscious hearing before the heavy silence swallowed all sound.

Then suddenly, without warning, a weird and startling sound split the uneasy calm and quiet of the lake! Jeb froze in his boat with his pole still in mid-air.

It was a high, twanging sound, completely alien to the swamp. Never had Jeb heard such a sound in the swamp before, and its impact on him was startling. For moments he stood frozen in his boat, wondering what to do, while the sound vibrated in the still, hot air like a plucked fiddle string, before gradually fading away. And all the time, Jeb's heart pounded and his mind raced wildly. What was it? Jeb hardly breathed as he strained to identify it, but the lake quickly became quiet again, and there was no answer. Only the deep, brooding stillness pushing in on him once more. Jeb remained motionless, listening with every nerve in his body, while the strange haunting sound vibrated over and over again in his mind.

Time passed and Emerald Lake remained completely surrounded by the deep silence. Not a frog croaked. Not a bird sang. But oddly enough, the quiet was not reassuring, for there was a tenseness about it that tugged at all his senses.

Then, the sound rent the air again! As high and piercing as before! It penetrated the whole region, vibrating mysteriously on shimmering waves of air. Completely baffled, Jeb let his troubled gaze sweep over the lake, the shoreline, and the island. But he saw nothing unusual. Then, taut and tense, he spun into action.

The sound had come from the end of the lake, and Jeb hurled his punt in that direction. He was so close to the shore that bushes and plants, jutting out over the banks, obscured his view as he frantically steered his punt toward the far end of the island. He had to search out the source of those strange and haunting sounds.

The air over the lake still quivered uneasily, but there was only silence now. More and more puzzled, Jeb strained his ears. Listening, listening. But all he heard were his own loud heartbeats, and all he saw was the dark shadow his small boat cast on the sunny water as it followed the curve of the shoreline. More and more shoreline slid past as Jeb became more and more tense and uneasy.

Then Jeb's boat rounded a thicket of bushes. Suddenly, a quick movement caught his eye! A furtive movement in the forest across the water from the island. Something was just disappearing into the dense growth beside the lake. Something that looked like the end of a boat. The impression was so strong that it registered with lightning speed on Jeb, and he stopped quickly, bewildered and afraid. A little breath of terror escaped him, and the fine hairs prickled along his sun-tanned arms. Struggling for self-control, Jeb told himself he was being over-imaginative. In the dim light and half shadows cast by the forest, ordinary things sometimes seemed strange and different, creating fantasies of fear. After all, it was only an impression. A shadowy image that had now vanished. Jeb's mind raced on. How could it have been a boat? This lake was known only to him. He alone had searched it out. He alone had discovered it. Only one narrow channel led to this isolated spot, and only he had passed that way.

Jeb told himself it was an animal. But what of those weird and startling sounds? No animal in the swamp ever made such sounds. With a sinking feeling, Jeb realized he could no longer ignore the creeping fear that threatened to engulf him. He could not ignore the chill feeling of strangeness and mystery that hung over the lake, penetrating to his very bones.

At last his mind had to accept that which, somehow, he had sensed from the beginning. Someone else had been here. Someone had made those strange sounds, had heard him coming, and had vanished. Vanished in a boat through a seemingly impenetrable curtain of jungle growth.

Deeply troubled, Jeb stared at the place where the boat had vanished. Many thoughts flashed through his mind. He knew that this swamp had, from long-ago times, been the haunt of the hunted. During the wars, it had been a refuge for runaway slaves and deserters. Moonshiners and outlaws had hidden in its jungles. But it was different now. Or was it? The question sent a shiver up his spine. Jeb was certain he knew all the people who lived along the borders of the swamp. And he knew their boats on sight. But that half shadow, that phantom boat of the swamp, was none he recognized. The day was warm, but Jeb suddenly felt cold.

He took himself sharply to task. He knew he should examine the passageway the strange boat had used, and he tried to force himself to move. But still he held back, listening, while his heart beat in slow, sickening thuds against his chest.

Then Mac growled and moved in the boat, and Jeb glanced down sharply at the dog he had forgotten. With growing apprehension, Jeb noticed that Mac's body was trembling, while his nostrils quivered with the strangeness of a scent he could not recognize. The dog's fear communicated itself to Jeb, and reason lost out to terror. Sheer terror of the unknown. In a moment all thoughts of exploring the island were forgotten. For the first time in his life, Jeb was afraid to stay in the swamp.

Spinning the boat about, he gave way to pure panic as he bent almost double in an effort to lengthen his strokes. The boat shot over the water. In a flash, he was out in the middle of the lake, heading in a straight line for the dark channel through which he had come, the channel that would also lead him back to the sunny prairie and home. His brain was spinning out of control, whirling and jostling with dark and frightening images, while he dug his pole violently into the water, pitting his muscle against the floor of the swamp.

Jeb made one last effort to pull himself together,

but he could not escape the sense of foreboding that hung over the lake. He could not forget the weird, high-pitched sounds. He could not blot out of his memory the sight of a phantom boat disappearing into the forest.

In haste, he dug his pole into the muddy bottom and pushed with all his strength. Time and again he bent forward, using longer and longer strokes to take him away. Even the run through the dark bay was preferable to the hidden danger that threatened here. For snakes and alligators were dangers he could see and understand, but Emerald Lake held something far more frightening—a terror of the unknown.

In a blur of exhaustion, he saw the vague shadow of a kingfisher perched in a tree beside the water. Grimly, he heard it rattling out its quick, loud notes, drumming him out of Emerald Lake. In shame, he realized that he was running away. In sorrow, he understood that the dream of the morning lay destroyed and dead, and the bright beauty of the island was now dulled and tarnished.

Still, nothing could have induced him to stay, and it was with great relief that he saw the dark entrance to the channel just ahead. In a frantic scramble, he pushed through the entrance, giving a half-glance over his shoulder at the lake behind him. His distorted vision saw only terror, shimmering like heat waves, over the dark, green waters of the lake.

With a sigh and a shudder, he turned his attention to the long, twisting run that lay ahead. Once more he threaded his way down the amber-colored ribbon of water that wound in and out among tall cypress trees.

Jeb was so intent on his poling that at first he barely noticed a white mist rising slowly from the water. It drifted up in little, quivering fingers. Then, gradually, it grew into great, swirling, white clouds that obscured the trunks of the trees and could not be ignored. With a sinking heart, Jeb saw the run that had seemed mysterious before, turn shrouded and ghostly. Unable to see the channel or steer a straight course, he raked his boat constantly against the sides of the run, ripping and scraping through foliage. Bushes and vines tore at his boat like living things. Still, he pushed recklessly on. He knew he was acting in panic, but he couldn't seem to help himself. As he plunged through the mist, reality ceased to exist, and Jeb made the rest of the trip through the forest in a blinding, white nightmare.

In a tight, flat voice, Jeb ordered Mac to stay low in the boat. Fervently, he prayed that the snakes he knew were still there would hear his boat scraping the sides of the run and disappear in time.

Everywhere, long streamers of moss moved to and fro, reaching out like cold, wet spider's webs. Time after time their clammy strands caught at his face, as if they were struggling to free themselves from the trees and join the mist ghosts from the water below.

On and on he went until, after an unbearably long time, he saw that the mist was slowly thinning. With a renewed burst of energy, and a final thrust of his pole, he sent his boat through the last cloud of fog; out into the brilliant sunshine of the surrounding, watery meadows. Relief flooded through him; it was like a release from a bad dream to emerge into the open swampland. Though he was nearly blinded by the bright sunshine and was near exhaustion, he began to relax for the first time since his strange adventure. Yet he felt moody and depressed, for he knew he had failed miserably by running away. He had left a part of his honor and integrity behind in the swamp. Already, he regretted his decision to leave the lake without searching the green bushes at the edge of the forest. Already, he knew that it had been wrong to leave the mystery unsolved. For he would never feel at ease in the swamp while some unknown terror lurked in its depths, or at ease with himself while he knew himself a coward.

With a sigh of remorse, Jeb set out across the watery prairie. To cover up his shortcomings, he promised himself that he would return to Emerald Lake after he had thought things through and had taken time to make some plans.

Out in the open swamp, he gradually felt safe and secure once more. Something of the nightmarish quality of his experience was worn away by the familiar sights and sounds of the meadowland. He watched the bees hovering over the golden bonnets, and listened to the birds and frogs singing the age-old symphony of the swamp. He concentrated on the countless alligators that sunned themselves on the mudbanks and rotten logs. He watched intently as they slid into the water when his boat approached, and he followed their strong bodies with his eyes as they plowed through the thick grasses and lilies.

Actually, Jeb had a special liking for and interest in the lumbering alligators, with their broad heads and long, rounded snouts. For he knew that they belonged in the swamp as he did. They were a part of the swamp, and it would have been a sorrier place without them.

Silently, he slid past one shining gator pool after another, watching the small, silvery fish that darted about in them. These were the pools where the alligators trapped their fish and hid during the nights. It was in just such a pool that Jeb had caught a baby alligator a year before. He thought affectionately now of that gator, which he had since trained as a pet. His face lighted as he realized that soon he would see Nod.

Jeb's family, like the other swamp people, lived in the flat, piney woods that bordered the great swamp. Their cabin sat on the edge of Broad Creek, which flowed like a watery highway right into the

Okefenokee.

Soon Jeb's boat neared Broad Creek and he quickened his pace, for now his home was not far away. Skillfully, he guided his punt to the place where the creek and the swamp merged, and left the gentle waters of the Okefenokee for the swifter-flowing waters of the creek.

4

POLING FORCIBLY AGAINST THE STRONG CURRENT
of the creek, Jeb headed for home, just a mile up-
stream. Now everything was familiar. The piney
woods on both sides of the creek crowded right
down to the water's edge. Somewhere in their cool
depths, a mockingbird welcomed him back with a
song of greeting. Even the creek waters made talking sounds that seemed like words of comfort to Jeb.
Mac, too, was eager to get home and shifted about
ever so slightly in the boat.

Before long Jeb saw, in the distance, the wide
clearing near his home with row after row of white
beehives shining in the sun. Next the neat garden

43

patch came into view. Then, towering above the gum tree, the great stone chimney that dominated the cabin's sharply slanting roof. And finally, the rest of the little, old, log house weathered and worn, but looking snug and secure within the confines of a split-rail fence. Although the air was still as Jeb approached, he could smell the faint odor of the noon meal cooking on Ma's big kitchen stove.

Continuing up the creek, Jeb's thoughts see-sawed back and forth in deciding whether or not he should tell Ma about the mysterious happenings at Emerald Lake. But the more he thought about how his story would sound, the more convinced he became that he could not tell about his strange adventure. He could not possibly describe the chill aura of mystery and fear that he had felt so strongly. How could anyone who had not actually heard those weird, vibrating sounds, understand their awesome impact? How could he ever make anyone feel the shock he had experienced when he saw a phantom boat disappearing into the thick, jungle-like growth of the forest? In truth, who would really believe he had actually seen a boat?

Jeb realized that Ma would wonder why he had come home so early, for he had told her not to expect him back until late afternoon. But somehow or other, he must try not to show that anything was wrong. Somehow, he must cope alone with the terrible knowledge that a mysterious boatsman was roaming about somewhere in the unexplored depths of the swamp. Somehow he would have to hide the fear that now made it almost impossible for him to return to the great swamp.

Suddenly, Jeb felt an overwhelming desire to be safe inside his own cabin where he would be in familiar surroundings and could talk to his mother again.

Throwing back his head, he yodeled loud and clear. Ma heard it and knew he was almost home. And Nod, his little pet alligator, heard it too. Sunning himself on the boat dock, Nod moved his small body ever so slightly and waited. Through half-closed eyes, he watched Jeb hitch his boat to the dock and jump out onto the weathered boards. Then he made a happy, clumsy move toward Jeb, who reached down and affectionately rubbed him on the hard skin between his eyes. As curious as a cat, Nod sniffed the bottom of the bucket that Jeb carried and, when he sensed that it contained fish, lumbered after Jeb all the way to the house. Then, as they neared the kitchen door, Nod opened his mouth wide and begged for food. But Jeb was so absorbed in his own thoughts that he completely ignored Nod. He was making a silent wish that Ma wouldn't ask him any questions. When he looked up, she was there, standing in the doorway smiling.

"Why, Jeb, you're back in time for dinner. It's a good thing I heated up the left-over chicken and dumplings."

For an instant, Jeb saw a question start to form

on her lips. But then her attention was drawn to his bucket of fish, and her eyes lighted up when she saw his large catch.

"You know, you're getting to be as good a fisherman as your father, Jeb. And it's a good thing, too, because we'll need all of them when Pa and Dr. Bowen get back from the swamp."

Jeb brightened and even boasted a little.

"Look, Ma, I'll bet this is the biggest one I've ever caught." He held up his largest fish for her to see.

Before long, the rich fragrance of chicken and dumplings, that permeated every cranny of the kitchen, reminded Jeb that he had eaten nothing since before dawn.

"Ma, I'm so hungry," Jeb pleaded, "could I eat first and clean the fish later?"

Before answering, Ma lifted the heavy lid from the iron kettle and touched a plump dumpling, gently, with the tines of her fork.

"Yes, Jeb," she said, "the dumplings are just right for eating. The fish can wait until after dinner."

Jeb sat down at the kitchen table and Ma sat across from him in her blue, gingham dress and heaped his plate high. While he ate, Jeb looked around him at all the familiar things of home. At the friendly cookstove with the copper kettle, singing away on its polished top. Then at the golden sunlight streaming through the kitchen window, spilling over the large pots of red geraniums and out onto the wide planks of the wooden floor. He was even conscious, for the first time, of the simple beauty of the brightly-colored rag rug that covered the entire center of the large kitchen. Every object in the room reflected a warmth that helped to dispel his dark thoughts, until he felt completely relaxed and at peace with himself.

After his first hunger was satisfied, Jeb began telling his mother about his morning in the swamp. Ma loved the swampland as much as Jeb; for she, too, had been born and reared in the Okefenokee.

Ma knew all about the wildlife, and she always enjoyed hearing Jeb's account of the flowers and animals he had seen. She drew in her breath a little when Jeb told her of his close call in the cypress bay with the cottonmouth, and nodded her head in approval when he described how he had finally frightened it out of the tree. Though Ma was used to the hundreds of snakes in the swamp and took them pretty much for granted, like all people in the region, she respected them and gave them a wide berth.

So far, Jeb had made no mention of the island and now, in the midst of his story, he hesitated. As he watched his mother's face, he considered again the possibility of telling her about the mysterious happenings. But the same doubts rushed through his mind once more. Wouldn't Ma think him scary and silly if he told her about the strange sounds he had heard? Could she possibly understand the strong feeling of mystery that surrounded the isolated spot? And what would she say if he told her he had seen the tail end of a boat disappearing into a tangled, vine-choked forest? Or that he had run away?

Even as he asked himself the questions, he knew the answers, so once again he decided to keep the mystery of the swamp to himself. It was not time to tell. Not yet.

Ma interrupted his thoughts, "There's a lot more chicken and dumplings, Jeb. With Pa away, you can eat your fill."

Jeb watched as Ma bent over the iron pot. When

she straightened, her face was flushed from the heat of the stove and she brushed back a wisp of rich, brown hair from her cheek. After she returned to the table with his plate, Jeb turned his eyes away from the direct look in her eyes. And, for an awkward second or two, there was a sharp little silence while he hunted for words to steer the conversation away from anything that might touch on the hidden island in the distant lake.

Jeb and Ma sat at the kitchen table and talked for a while longer, even after the meal was finished. But finally, Ma got up.

"Jeb, I've got to get started. I promised Mattie I'd help with quilting this afternoon."

Tying on her sunbonnet as she went out the door, Ma reminded him, "Don't forget to clean those fish and put them in the springhouse to cool. And Jeb, if your father and Dr. Bowen get home before I do, be sure and invite Dr. Bowen to stay for supper."

Jeb promised. Then he watched Ma's brisk step going along the path through the piney woods to Miss Mattie's house, two miles up the creek. He knew that Ma wanted to be sure she got to see everything that Pa and Dr. Bowen brought home with them from the swamp.

Dr. Bowen worked for the museum in Raleigh. Pa's interest in the swamp had made him more than willing to guide Dr. Bowen to Pine Island where they would excavate one of the ancient Indian burial mounds, in which the archaeologist hoped to find

Indian relics to take back to the museum. Pa had gone on such trips before.

Jeb again mulled over the odd coincidence of seeing a mound on the hidden island at a time when his father had become involved in excavating a similar one at a far-distant site. He tried to imagine what kind of artifacts the men would bring home with them. In the past others had found pottery, tools, and weapons, different from and far superior to those made by any other known tribe of Indians. Yet the story of the mound-builders seemed lost forever in the ancient history of the great swamp. Still thinking about the mound-builders, Jeb picked up his bucket of fish and started for the creek to clean them.

Nod was still at the kitchen door peering in, and it amused Jeb to see his black, shiny eyes watching so intently through the screen. When Nod saw Jeb, he opened his jaws and yawned slowly. Nod was still a baby gator; in the years to come, he could easily grow to be twelve or thirteen feet long. Like all gators, he did not like to be far away from water, so he happily followed Jeb back to the creek.

At the creek, Jeb sat on the pier and Nod floated in the water almost submerged; together they played a familiar game. Cutting off a fish head, Jeb tossed it to Nod. Then the water came alive, boiling and churning, as Nod's powerful little tail slapped the water and his agile body thrashed about until he was within reach of the fish head, which he caught in his open jaws.

Nod greedily swallowed several morsels of fish. Then, when he became full, he began diving beneath the water with other fish tidbits that Jeb threw to him. He was storing them in his cave in the muddy bank far below the surface of the creek. There the food would soften and he could eat it when he was hungry again.

Jeb had just finished cleaning the last of his fish when he heard the sound of an automobile coming over the narrow, dirt road that led from the main highway to the cabin. Hurriedly, he filled his bucket with water to cover the fish and ran, with Mac at his heels, to meet the car, which was just stopping under the large gum tree near the cabin door.

Jeb recognized Mr. Garret and Mr. Fry as soon as they got out of their car with their fishing equipment. Pa had guided them into the swamp to the fishing spot almost every day for a week in the early part of the spring.

"Is your father home, Jeb?" asked Mr. Garret.

Jeb shook his head. "He's in the swamp now, and won't be back until supper time."

Mr. Fry arranged his fishing gear more securely in his arms.

"We were just passing through and thought we'd take the afternoon off for fishing. We'd certainly hate to go through the best fishing country in the south without wetting our lines. Do you have a boat we can use, Jeb?"

Jeb hesitated. There were two boats at the dock.

His little punt and a larger boat. Jeb knew that his father would never let an outsider go into the swamp without a guide and now, with his father away, he knew he should refuse to lend them a boat unless he went along. But, for the first time in his life, Jeb didn't want to go into the swamp, and especially not to the fishing spot. The start of the run was too close to and too much like the one he had fled this morning. Just the thought of returning brought back all the memories of those strange sounds and shadows.

Finally, however, Jeb forced himself to offer, "I'll let you have a boat, but I'll have to guide you in."

Mr. Garret shrugged, "That won't be necessary. We know the way well enough. After all, we went into the swamp almost every day for a week just last spring."

Now was the time for Jeb to insist that he go along. And he knew that if he did insist, the men would have to agree to use him as a guide. But Jeb instead kept silent. He knew only too well the dangers that outsiders risked in going into the Okefenokee without a swamp man to show them the way. Swamp men could find their way about by remembering where a particularly odd-shaped cypress tree stood. Or they used the nests of eagles or the roosts of buzzards to guide them. Outsiders didn't know how to recognize these "lighthouses" of the swamp, and they often became hopelessly lost.

With all these thoughts flashing through his

head, Jeb still kept silent.

The men were in a hurry to get started, but they had driven without stopping for lunch.

"Is your mother home, Jeb?" asked Mr. Fry. "Maybe she would fix us a few sandwiches to take with us. And we'd like a jug of that good spring water."

Jeb shook his head again. "Ma's out, but I'll fix you some food."

He went into the house and packed several sandwiches in a paper sack, grateful for some time to think of a way to keep the men from going into the swamp. Then he filled a glass jar with water from the spring. Carefully, he put on a tight-fitting cap. Outsiders shied away from drinking the swamp water, though the swamp folk knew that it was pure and sweet, and just took along a cup to use in dipping it up. After the lunch was packed, Jeb carried it to the men who were down at the creek, already waiting in the larger boat.

Jeb's head whirled and raced, but he could think of no way to keep the men from leaving. So, with grave misgivings, he handed the food and water down to them. Standing there alone, unable to handle the situation, Jeb missed his father terribly.

He tried to get up the courage to either stop the men or go with them as Mr. Fry began to pole slowly down the stream. But instead he stood silently by and watched the boat go down the creek, around the bend, and out of sight.

5

Left alone, Jeb brooded and tried to forget the two men he had allowed to go into the swamp without a guide. First, he wandered aimlessly down past the rows of beehives, where the air hummed with hundreds of bees carrying nectar from the tupelo trees of the swamp. Then, remembering that he had left his bucket of fish on the ground under the gum tree, he ran back, got it, and quickly stored it in the cool darkness of the springhouse.

Still in a bleak mood, he went into the house and got a small piece of soft, white, pine wood that he had already started carving into the figure of a bear.

Finally, he settled down on the boat dock, with

Mac curled up asleep beside him. There, in the warmth of the afternoon sun, he took out his new pocketknife and continued his carving. Presently, a pile of white, pine chips lay on the dock and an amazingly lifelike figure of a black bear took shape. Jeb, who was familiar with most of the common wild animals of the swamp, liked to carve their likenesses in wood. Soon the bear would be added to the small wooden fox and wildcat already on Ma's window shelf.

Though Jeb whittled away furiously, and tried to focus all his thoughts on the task at hand, his mind kept returning to the mysterious happenings in the swamp, or jumped to his newer worry over the two fishermen. Weary and depressed, he realized that the brief peace he had felt in the warm, bright kitchen while he was talking to Ma was illusory. For him there could be no real peace of mind until he could once more venture into the swamp without fear, and the two outsiders had returned safely.

The afternoon hours passed slowly on, and the sun sank lower and lower in the sky. Soon Ma would be coming back through the woods to start supper. And Pa and Dr. Bowen would be bringing home the treasures they had excavated from the Indian burial mound. Jeb, who had been beside himself with curiosity all week, who had thought he couldn't wait until he saw with his own eyes the things the ancient Indians had hidden in the mound, now dreaded to think of the men's return. For he could imagine what his father would say when he found out that Jeb had let two outsiders go into the great swamp without a guide.

Then a pleasanter thought occurred to him. Perhaps the two fishermen would return before Pa did. They had only wanted to fish a few hours. Jeb felt a great surge of hope, for if the fishermen returned first, Pa would never need to know what he had done.

As if in keeping with his thoughts, a boat appeared far down the creek. It was too far away for Jeb to tell who was in it, and several minutes passed before he could make out the tall, wiry figure of his father poling up the stream.

With mixed feelings, Jeb waved as Pa guided the boat close to the dock. Then Pa caught hold of the pilings, deftly looped a rope around a pole to secure the boat, and leaped onto the dock.

Pa's excitement was apparent in the way he briefly squeezed Jeb's shoulder.

"Wait till you see what we found, Jeb. The pottery is still whole, and there's an enormous wooden bow. This must've been the grave of a great hunter or an honored warrior."

It was estimated that there were about twenty good-sized islands scattered throughout the great swamp, and many had mounds on them, or probably did. Not all had been explored. Dr. Bowen had chosen a mound on Pine Island to excavate because of its large size and careful construction. This indicated that it sheltered an important member of the mound-builder tribe.

Jeb knew that the carefully wrapped parcels that Dr. Bowen handed gingerly to Pa from the boat contained the best of the relics. He guessed that many more boat trips would be made before all of the excavated treasures were brought out of the swamp.

Jeb felt a little hurt that neither Dr. Bowen nor Pa asked him to help carry the artifacts to the cabin. They only asked his aid in holding open the cabin door while they carried the relics to the third bedroom and laid them carefully out on the large, wooden table that Pa sometimes used as a desk.

With the utmost care and gentleness, Dr. Bowen unwrapped a small pottery bowl, and Jeb felt a surge of excitement as he saw the exquisite figures of tiny, painted deer leaping around the lip of the bowl. Several more pieces of pottery were unwrapped and displayed, and then Dr. Bowen showed Jeb a tiny, curved fishhook, carved delicately from bone. Jeb felt a strange sense of kinship with some long-ago Indian fisherman who had used this fine, bone hook in the swamp waters.

But the bone fishhook also brought Jeb's mind back to the fact that the two outsiders had not yet returned. Jeb glanced quickly out the bedroom window and was relieved to see that the sky was still light. Yet, a feeling of uneasiness remained; he hoped that the two fishermen were on their way out of the swamp. Jeb knew that his father had been so excited over the Indian relics, that he hadn't even noticed that a boat was missing from the dock.

Then Ma came breathlessly into the cabin. She had been hurrying so she wouldn't miss any of the excitement. Although it was almost seven o'clock, no one thought much about supper as Dr. Bowen continued to unwrap, and hold up for inspection, one fascinating artifact after another. There was more pottery, a stone ax, and several cutting tools of stone. And all of the relics showed the same high degree of artistry and skilled craftsmanship.

Finally Dr. Bowen held up the two prize pieces of them all—a magnificent, heavy, wooden bow and one perfect arrow. The bow showed the many hours of work and loving care that the unknown warrior had spent in shaping and polishing the beautiful wood, and Jeb's eye was drawn to one end where a piece of sinew bowstring was still attached.

Then, Dr. Bowen pointed to the slender arrow that had been found next to the bow in the grave and explained, "A live hunter needed a lot of arrows, but a dead hunter needed only one. On that arrow his spirit rode, straight and true, to the Great Warrior God."

When Dr. Bowen noticed Jeb's intense interest, he handed the arrow to him to examine more closely. Jeb turned the arrow around slowly in his hands, noticing every detail. The shaft was smooth and highly polished, like the wood of the bow, and fastened to one end with sinew was a long, slender,

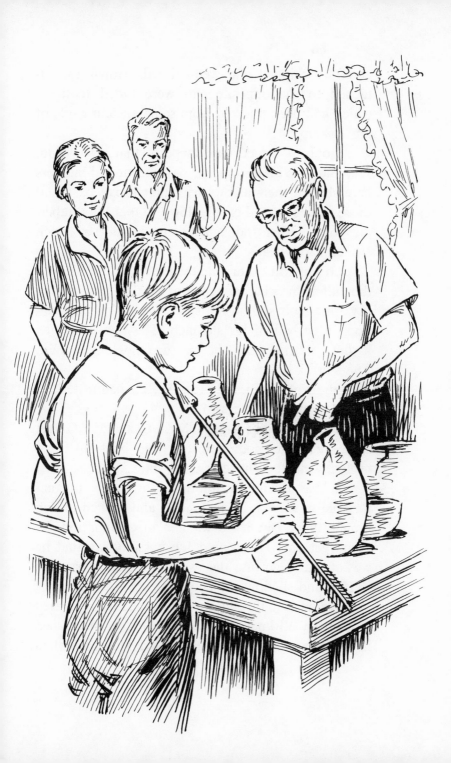

carefully chipped and well-pointed arrowhead. Attached to the opposite end were faded feathers, which had been plucked from some ancient eagle of the swamp.

"Look at the curious shape of that arrowhead, Jeb," said the archaeologist. "All the mound-builders here shaped their arrows with that broad base and those sides slimmed evenly down to that long, tapering point. Other Indian tribes made shorter, much cruder arrowheads."

Jeb was still examining the arrow, lost in his own thoughts, when he happened to look up and see the sky. One look and he felt sick, with a slow fear settling in the pit of his stomach. The two fishermen hadn't returned, and the sky was no longer light. The sun had set.

Jeb turned then and faced Pa. He had to tell his father what had happened.

"Pa," he said haltingly, "Mr. Fry and Mr. Garret came right after noon and took your boat out to go fishing. And they haven't come back."

Pa was obviously stunned. For a few uncomfortable moments, he just stood and looked at Jeb thoughtfully. But if he wanted an explanation, he didn't get one. How could Jeb possibly say that he was afraid to go into the swamp because of strange sounds and shadows he couldn't explain?

Pa and Dr. Bowen and Ma exchanged worried looks; they all knew the dangers the two men faced if they were lost. Pa's forehead wrinkled and when he spoke, he looked very tired. "There's no point in trying to find them at night. We'll have to wait until dawn tomorrow and get an early start. I only hope they haven't strayed too far from the regular boat lanes. If they have, we'll need help."

Even though Dr. Bowen stayed for supper, it was a rather unhappy and silent meal. Everyone was listening and hoping for sounds that would indicate the two men were returning.

Before long Pa suggested that everyone turn in early so they could start searching at daybreak. Dr. Bowen agreed to stay overnight and help out in the morning, and Ma hurried to get his bed ready in the spare room. Then, after a curt good-night nod from Pa, Jeb went slowly off to bed, his sense of guilt further burdened by the knowledge that now Pa and Dr. Bowen wouldn't be able to return to Pine Island the next day as they had planned.

That night, though Jeb was weary and exhausted, he found it hard to go to sleep. Noises from the great Okefenokee floated in: the hoot of an owl; the hollow, booming bellows from the big, bull alligators; and once the piercing screams of a wildcat.

Then, when he finally drifted off, it was to toss and turn in a state halfway between wakefulness and sleep. Little uneasy snatches of sleep crowded with weird echoes of sound, and distorted by nightmares of a phantom boatsman; and two men poling endlessly on and on, lost and alone in the dark, mysterious swamp.

6

THE ENTIRE HOUSEHOLD WAS UP BEFORE DAWN.
Slipping out of bed, Jeb padded across the floor in
his bare feet and stared dejectedly out the window,
his eyes probing the great, bulky darkness that still
gripped the piney woods and the vast swampland
beyond. All sorts of emotions churned within him
as he watched the last dark of morning thin to a soft,
pale gray. Shame and humiliation still gnawed at
him. But though he yearned to set things straight,
he still shrank from the thought of going back into
the swamp.

Over breakfast Pa outlined his plans for the
search. Dr. Bowen, who was a stranger to the swamp

and wouldn't be much help during a search, was to drive into town and alert the men there to the situation.

Then Pa turned to Jeb and continued, "We'll go as far as the edge of the big forest together. Then, I'll go down the boat run to the fishing spot. If I don't find them there, I'll look through the cypress bay to the north. You hunt down farther south, just in case they went too far in that direction."

Jeb could only stare at his father, stunned. For the region to the south contained the waterway that led to Emerald Lake. He gave a little shudder at the thought of going back there by himself. It had never occurred to him that he and his father would go into the swamp alone to hunt for the men. He had expected Pa to call on the men in town immediately for help.

Now he realized that only if the two fishermen had gone into the unexplored parts of the swamp would a large search party be used.

Jeb had seen such large search parties operate in the past, and he knew the time and trouble involved. He recalled seeing the men split up into many teams, each yodeling a special call so others could keep track of them in the wild, uncharted regions of the swamp. The "hollers," as the swamp people called these yodels, had been used for years to communicate back and forth across long distances.

Even though Jeb well understood Pa's reluctance to organize a group like this until he and Jeb had searched the logical places, his anxiety made him blurt out, "Shouldn't we get the men from town first?"

As soon as Jeb had spoken, he was sorry. First his father looked puzzled at his reluctance to go into the swamp. Then he spoke curtly, explaining that it was their responsibility to try to find the outsiders before inconveniencing a large number of other men. The look and the words were a rebuke that Jeb knew he deserved and he dropped his eyes quickly as the blood rushed to his face.

Pa finished giving them their final instructions. They were all to meet back at the cabin by noon. If Jeb or Pa hadn't found the fishermen by then, the searchers from town would be called out.

Ma had scarcely begun to clear off the breakfast dishes when Dr. Bowen left in his pick-up truck for town. Then Pa and Jeb, each in his own boat, poled swiftly and silently down the creek towards the great swamp. Even Mac seemed to sense the seriousness of the situation and he barely moved as Jeb trailed behind his father, down the creek and out over the open water of the prairies.

Together they continued on through long stretches of open prairie, flecked with the gold of a thousand lilies. On and on, steering a course that led them across the grasslands towards the brooding, cypress forest that loomed thick and black against the far horizon. At last, they left the wide grasslands and passed by way of an old alligator trail, through the marsh of brown cattails, right to the border of the dark bay. Shortly afterwards, Pa signaled to Jeb that he was turning into the boat run that led to the fishing spot.

"Good luck, Jeb," he called. "Be careful."

Now without Pa, Jeb's unreasonable fear of the swamp returned. For a moment, he gazed back at the dark entrance into the forest through which his father had disappeared, remembering the countless times he had taken that same channel, either alone or with Pa. Like most of the runs in this region of the bay, it was shadowy, dim, and sometimes narrow. But it differed in one important way. Instead of twisting sharply backwards, forwards, and sideways, in all directions, it wound its way gently eastward, straight into the cypress jungle.

Jeb knew that Pa would pole down the run until it widened into a small lake that was alive with pickerel, bass, bream, and catfish. This was the fishing spot, and, because of its easy access, Pa often guided outsiders into it. Now he would go to see if the men were still there. But Jeb could think of no reason why they should have stayed there overnight, unless they had run into some kind of trouble. After all, the trip from the fishing spot back to the prairie was as uncomplicated as the way in. And even if they had decided to go farther down the run, past the place where it widened and deepened into the lake, they would have been forced, sooner or later, to return. For the channel beyond the lake gradually became so narrow and plant-clogged, that it defied all passage.

Jeb frowned as he tried to imagine what had happened to the fishermen. He was sure that after leaving Broad Creek they would have been able to find their way across the open prairie to the great forest.

But he suspected that they might not have been able to locate the old alligator trail that led, through the prairie reeds and cattails, almost directly to the entrance-way to the fishing lake run. If that were the case, the fishermen would have reached the cypress bay either too far above or below the fishing channel, and entered the forest at the wrong place.

It was impossible to guess where the men might be, but if they had gone too far south, it was up to him to try to find them. There was no use trying to avoid the inevitable. He had to face his problem now.

So Jeb threw back his shoulders and continued south along the border of the shadowy cypress bay. At the first boat run below the fishing channel, he turned into the forest and began his search in earnest. After following the run along a direct course for a fair distance, it split abruptly into two channels, with one going left and the other right. Jeb traced these out and then returned to the edge of the forest. From there, he progressed steadily south, investigating one dim, sharply-twisting watercourse after another. Slowly and thoroughly, he continued his search, wandering in and out of numerous bordering runs, carefully scanning his surroundings for a clue that might indicate that the fishermen had passed that way. Occasionally, he yodeled into the dark depths of the bay, but there was no answer.

And always, almost without realizing it, Jeb kept track of his passage. There were the twin, wooded islets where the night herons roosted; and there was the tree where the wood ducks lived. Even bushes and clumps of vegetation were landmarks to his practiced eyes. But certainly no stranger could ever hope to find his way about if he followed these tortuous waterways very far, for they usually doubled back and forth, interlacing among themselves, in such a way, as to form a confusing network of runs. Without knowing the landmarks in such a region, a man's sense of direction would be easily confused.

But would the fishermen have been foolish enough to take channels like these—that were so obviously wrong ones—far enough into the forest to actually lose their way? This was the question that Jeb had been asking himself over and over again all during the search. In his opinion, it was not reasonable to suppose they would, for they knew the kind of channel to take to the fishing spot. They had gone down that gently twisting boat run with Pa for a whole week in the spring. They would have wanted to take an eastbound channel. One that wound its way with gently twisting turns, straight into the forest.

Jeb recalled how determined the men had been

to get to the fishing spot. Suppose they had tried channel after channel, returning to the edge of the forest each time after finding a run to be the wrong one. If searching southward, where might they eventually have found a channel enough like the one to the fishing lake to have taken it? One that went straight eastward into the forest with a few gently twisting turns? As Jeb formed the silent question in his mind, he had to admit what was really bothering him. For he knew which boat run it was. It was the boat run to Emerald Lake, and it was the fear of returning there that haunted him.

Up until now, he had not let himself really think about how similar the fishing lake run was to the first part of the run to Emerald Lake. And, of course, the two runs were not a great distance apart. If he had found the run to Emerald Lake that afternoon after leaving the fishing spot to look for eagles, it was entirely possible that the two men might have found it while hunting for the waterway to the fishing spot. No run between the two would have seemed right to the men, he was sure.

But the run to Emerald Lake did not widen out into a fishing lake at the proper distance. After a time, the men would have realized that it went on and on into the forest. They would have known it was not the run they wanted.

Despite this reasoning, Jeb's heart began to pound, and he thought wildly of abandoning the hunt. His courage returned only when he realized that there was a fifty-fifty chance the men had never even come south. It was just as likely that they were in the region where Pa was searching.

He was also heartened by the thought that even if the men had come south, discovered the run, and taken it all the way to Emerald Lake, there would be no need for him to go in after them. How could they possibly have gotten lost there? With only one obvious channel leading into and out of the lake, they could easily have found their way back to the prairie.

Somewhat reassured, he struck off into another meandering, senseless waterway. And, now that his self-confidence had reasserted itself, he couldn't help but think how much trouble he would have saved if he had managed to overcome his fear yesterday and had gone with the fishermen into the swamp.

After a while, he emerged from the bay, skirted the forest a bit, and eventually found himself approaching the old, dead cypress that marked the boat run to Emerald Lake.

Despite the grim memories the run held for him, he knew he had to check the entrance for possible clues, and call a time or two into the bay at that spot.

Slowing his boat at the entrance, he eased his way in. Then, he ventured a little ways farther, yodeled down the dark stretch of the run and was greatly relieved when he heard no answer. Satisfied that the men had not been there, Jeb had started to leave, when a small, white object, caught on the knee of a cypress tree just inside and to the left of the entrance attracted his attention. He started to ignore it, but somehow or other it nagged at his thoughts. So he poked it with his pole and realized that it was a piece of cloth tied to the tree, the kind of cloth a fisherman might have in his tackle box.

Almost instantly, he knew what had happened. The fishermen had come to the run, had decided it was the right one at last, and had left a marker for themselves to indicate the end of the run when they came back.

Just as Pa had felt it necessary to check out the fishing lake, so Jeb now felt duty-bound to go into Emerald Lake. For while it seemed unlikely that the men would have lost their way on the one channel that led into and out of the region, there were other dangers they might have faced. In a flash, he recalled the strange current that, at one point along the run, had drawn his boat off to the left. Then, there were the snakes, a threat in any cypress bay but uncommonly thick in this long-hidden run. The danger that overshadowed all the rest, of course, revolved around the mysterious boatsman, the furtive boatsman, who might still be lurking somewhere at the lake.

As fearful and reluctant as Jeb was to enter the dark channel, he knew he had no choice. A lifetime of responsibility committed him to it. Yesterday, panic and fear had forced him to leave Emerald Lake. Today, a desperate hope of finding the fishermen compelled him to return. Jeb started the long journey down the gently twisting run.

In some curious manner, the horror he had felt so keenly yesterday was dulled by his bold action. Feeling strangely detached from his recent fears, he threaded his boat around the bottle-shaped trunks of the cypress and carefully skirted their knees and dark roots. Numbly, and almost without emotion, he maneuvered his boat on and on, while Mac sat in the bow of the punt, accepting as commonplace this third trip through the snake-infested channel.

Jeb lost all track of time as he sped on, but when he reached the sharp bend in the run, he realized how far he had come.

He slowly rounded the bend, skillfully dodging the thorny branches that extended out across the channel. Then he heard it! A quick rustling sound in the bushes beside the run. Immediately he stiffened and his heart gave a sudden lurch. Quickly, he checked his speed, while his head jerked sideways in the direction of the sound. Instantly, his detachment and apathy were stripped away, and panic again washed over him. In the moments of stillness that followed, his stomach muscles tightened and he became as alert as a fox.

Paralyzed with fear, he could do nothing but remain quiet and listen. At one time, he thought he saw one of the bushes bend just a little, but he really couldn't be sure that anything had moved. Still he waited.

Then he heard it again. A soft scraping and then

a rustling noise, closer this time and louder. He waited, not moving a muscle, for he fully expected to see the phantom boatsman enter the channel from a plant-clogged side-run.

Then something exploded out of the forest, followed quickly by a second object. Recovering from the sudden noise and motion, Jeb saw what had happened. A sleek otter, in a playful mood, had plunged recklessly down a slippery mudbank into the water, followed closely by a companion. Jeb almost laughed aloud in relief. Then, while his rapidly beating heart slowed down to normal, he watched the antics of the frisky pair of little clowns. Swift and agile swimmers, they raced and frolicked about in the water; then they scrambled back onto the mudbank to repeat the game, again and again. Mac was fascinated by the little otters and whined in excitement till Jeb cautioned him to quiet down.

Once more, Jeb continued on his way, but now he had become so wary that he navigated the rest of the run to the lake with the caution of a wild animal who senses danger but does not know from what direction it might come.

When Jeb reached the entrance to Emerald Lake, he paused while he carefully scrutinized the water and the surrounding forest. Then he looked across the lake to the green island at its far end. Everything seemed quiet and peaceful, yet he could not shake off his apprehension. Once more, his eyes searched the lake and forest and island, but he saw

no sign of the two men. Now he wondered if the clue
he had found was a false alarm. Certainly, he had
seen no evidence along the way to indicate that the
men had actually navigated the channel, and there
was no indication of their presence at the lake. Perhaps he had made the unwanted trip in vain.

Still, before he could leave, he had to make sure
the fishermen were nowhere in the region. He knew
he had to make a thorough search, yet he hesitated
before actually entering the lake, for he still half
expected to hear those strange sounds vibrating on
the still, hot air. His fingernails bit into the sweaty
palms of his hands as he drifted slowly and cautiously out onto the open water, but he forced himself to go on. Gradually, however, the incessant
croaking of frogs and the singing of birds, reminded
him that he need be afraid on the lake only when it
was excessively quiet, when someone or something
had caused the animals to cease their usual calls.

Once out on the water, Jeb quickly planned his
strategy. He would steer clear of both forest and
island, for by staying in the middle of the lake, he
could get a good view of both places, and would
be immediately aware of any sudden danger that
might arise.

So he kept a center course down the lake, scanning first the forest and then the island. But nothing
moved along their borders. On he went, over water
so clear and glassy that the sunlight, striking down
through its crystal depths, illuminated a large school

of tiny fish darting about far below. On and on, always searching for a telltale movement along the forest or island. But all that moved was a little brown duck flying across the lake scarcely a body width away from a mirror-like surface that reflected every detail of the smooth, rhythmic flight.

Although the search seemed utterly hopeless now to Jeb, he decided to stop his boat and call out. Through cupped hands, he yodeled first towards the forest region and then in the direction of the island. The sound, in the clear, early morning air, carried great distances across the still waters.

He had just started to call again, when he heard excited, garbled cries coming from the direction of the island. Jeb drew in his breath in amazement. Although this was what he had hoped for, he was actually startled to hear an answer. At length, the sounds became more distinct.

"Over here! We're over here on the island."

Jeb was astonished. He had really found the lost fishermen. Quickly, he poled towards the island, calling over and over again, while the men's answering voices guided him on.

Carefully, he edged his way in and out of the thick, plant growth that cluttered the shoreline. Finally he was almost opposite that part of the thick forest into which he had seen the mysterious boatsman disappear. Hurriedly, he poled past the place, all the while searching the thick, jungle curtain with his eyes for a sign of a channel, but again none was visible. The very sight of this part of the forest made Jeb uneasy and he was glad when, a short distance away, he saw the fishermen's boat anchored in a shallow cove of the island. Quickly, he guided his own boat alongside it.

At last, Jeb was about to set foot on the island that he had discovered, but, ironically, he was not the first to land there. The two fishermen, and perhaps someone else, had beaten him to it. As Jeb prepared to leave his boat, he noticed that this island, like most of the others in the swamp, had thick bogs of vegetation close to its shoreline. Though this area looked firm and hard, it was really unstable and insecure ground, composed of mucky layers of leaf-mold and moss held loosely together by the roots of the plants that grew there. Carefully, Jeb stepped out of his boat onto the spongy ground. Immediately, the earth trembled beneath his weight, and trees and bushes growing close to the water's edge swayed crazily as he walked gingerly to higher, firmer ground. As he slowly picked his way, with Mac close at his heels, Jeb couldn't help but think that the swamp had been rightly named. For Okefenokee is an Indian word meaning "land of the trembling earth."

The two men came rushing to meet him, smiling and waving. As they approached, he looked them over carefully, but could see nothing wrong. The fishermen saw the puzzled expression on his face and looked embarrassed.

"Jeb," Mr. Garret said, "we looked a couple of hours yesterday for the channel to the fishing lake. When we finally found the one coming here we thought we had it. But it kept going farther and farther into the forest, until even we knew that we were on the wrong one. By then, it was so late we knew we'd never get to the fishing spot. And we hated to waste a whole afternoon without even getting our lines wet, so we just kept going, hoping the run would widen out into a pond. We had just about decided to turn around and go back, when we came to a sharp bend and saw that the channel got wider and lighter. That made us think we might come to a lake, so we followed it through to the end, hoping we'd still get in some fishing."

Jeb didn't reply right away, for he was lost in his own thoughts. He had guessed right. The men had searched for an eastbound channel. But while he had spent only two hours or so checking out the runs and going into Emerald Lake, the fishermen, who were inexperienced boatsmen and unfamiliar with the region, must have spent two or three times as long. It must have been late in the afternoon when they finally reached here.

When Jeb broke away from his thoughts, he noticed that both men were smiling at him rather sheepishly, and Jeb grinned back. For all three were remembering the day before when the fishermen had rejected Jeb's offer to guide them. But only Jeb knew the truth—that he had really been glad they hadn't wanted him along.

Mr. Garret continued, "The fishing was good, but we'd only fished an hour or so when it began to get dark. Jeb, we tried to get up courage to go back down the run, but it was pitch black in there after the sun set. And that run! Those snakes all the way. Well, it was bad enough in the daytime. We couldn't face it at night.

"We knew people would worry if we didn't get back on time, but between the run and not knowing our way around the swamp very well, we thought we'd better spend the night on the island and find our way out in the morning."

While Jeb was digesting the fact that the men hadn't been hurt or lost, but had only run out of time, Mr. Garrett reached out and shook his hand. "If we hadn't stopped to cook and eat our fish this morning, we'd have been out of here by now and saved you the long trip in. We know we've been a lot of trouble and we're sorry."

Mr. Fry nodded in agreement. "Next time we go fishing in the Okefenokee, we'll have sense enough to take a guide."

The two unshaven men were ready to leave. Jeb could see that. And now they headed for their boat. But Jeb glanced back at their camp site, where they had spent the night. He saw that they had slept on the sandy ground and had obviously had a fire, for a few wisps of smoke still rose from the hot coals.

"I'll be with you in a few minutes," Jeb said.

"I'd better put out the last of your fire."

The men had dug a shallow depression in the sandy soil and had surrounded it with rocks to make a fireplace for cooking their fish. The remains of several fish were nearby.

Jeb had turned to scoop some sandy soil on the embers, when he saw a bright red object on the trunk of a nearby tree, about two and a half feet above the ground. At first he thought it was a woodpecker's red cap. But it didn't move. He looked again, and then ran to it quickly.

A little tremor of excitement ran up and down his spine as he examined the object carefully. For several seconds, he stood there, transfixed. In this wilderness, on this secluded island, he was staring at an arrow. The red, which he had noticed first, belonged to the feathers fastened to one end. The other end was embedded about a half inch in the soft bark of the pine tree. Jeb's hands were shaking as he carefully grasped the polished shaft and moved the arrow back and forth to loosen it. After tugging gently several times, the arrow came free.

Turning the arrow around and around in his hands, he felt the smooth wood of the shaft, then the stiff spring of the feathers. He examined the arrowhead. It was all there in his hands, and yet he found it hard to believe. For he could see in his memory another arrow that was easily the twin to this one.

That other arrow was in his cabin, in the third bedroom, on the great, wooden table. It was spread out among the collection of Indian relics that had come from the burial mound; and it was hundreds of years old.

Although that ancient arrow could easily be this one's twin, there was one tremendous difference. The one at home had feathers that had faded to gray with age. But this one had fresh, bright, red feathers to guide it along its path. Jeb could hear again Dr. Bowen's explanation of the oddly shaped arrowhead, "Look at the curious shape of the arrowhead, Jeb. All the mound-builders here shaped their arrowheads like this."

All the mound-builders—but the mound-builders were gone from the swamp. They had disappeared long ago. What, then, was a mound-builder's arrow doing here on this remote and hidden island?

A hundred unanswered questions rushed through Jeb's mind, and in the excitement of his discovery he completely forgot the two men waiting for him. But their sharp call brought his attention back to them. Swiftly and very, very carefully, he put the arrow under his shirt and buttoned it back up again. There was no more time to speculate, but he knew what he would do. He would take the arrow back and then, when he was alone, he would put it side by side with the ancient arrow in the cabin. He would carefully compare the two.

It was on the long trip back home, with the two

outsiders following close behind him, that Jeb made his plans. He would return to explore the island as soon as he could, possibly the next day. And he would come alone. Though his fear remained, his curiosity was greater. Perhaps, he would find more arrows. Perhaps he might even find out more about the high, twanging noises he had heard. But in the back of his mind, Jeb suspected he already knew the answer. For a bowstring snapping and then vibrating in the still, summer air would make such a sound.

Jeb was scarcely aware of the winding journey back through the cypress forest, for his thoughts were all on the arrow under his shirt and the deep mystery that surrounded it. Instinctively, he guided the boat down the channel, and, like one in a dream, answered the men when they spoke to him.

The time spent on the journey home seemed interminably long. And the time he would have to spend at home waiting until he could venture out at dawn the next day, already weighed heavily upon him.

When Jeb and the two fishermen finally reached the cabin, Dr. Bowen, who had returned a few minutes earlier, ran out to welcome them. Time passed quickly until Pa, who had searched the forest up to the last possible moment, pulled up at the dock just a little before noon. When he saw the smiling group waiting for him on shore, he hurried from his boat to take part in the reunion.

Lunch that day was a far happier meal than supper had been the night before. The great, round, kitchen table was heaped high with sausage, home-made bread, and honey. A relaxed, pleasant atmosphere pervaded the sunny kitchen as Ma bustled happily from stove to table making sure everyone got enough to eat. The tired look had gone from Pa's face, and Dr. Bowen looked relieved, too.

Pa was pleased that the two men were back, but it was obvious he was still wondering why in the world Jeb had let them go into the swamp alone. Jeb could tell what Pa was thinking, for whenever he glanced up, he found Pa looking at him in a questioning way.

Certainly, Jeb was relieved that the two men were safe. But he was far from calm, for his imagination soared whenever he thought of the island and the newly found arrow, now in a box in his bureau drawer. He had not yet had time to compare it with the one he knew to be its twin, and he was impatient for the meal to end. But he ate what Ma passed him and helped Mr. Fry and Mr. Garret tell the story of their meeting on the island.

Before the meal was over, Jeb mentioned casually that he would probably get up early the next morning to explore the island where he had found the two men.

Then turning to Pa he said, "I might decide to stay in the swamp for a few days. Camping out will be good training for Mac."

Though neither said any more on the subject, Pa knew now that Jeb was no longer reluctant to go into the swamp.

Presently, Jeb was able to slip off alone. First, he got the newly found arrow, and then he took it to the third bedroom, where he laid it beside the ancient arrow from the Indian mound. If Jeb had had any doubts at all that the two were alike, he had none as he stood silently comparing the two. But he was more puzzled than ever before. Though finding the arrow had removed some of the uncertainty and terror from the sounds Jeb had heard at Emerald Lake, it had only compounded the mystery. Why would anyone use an arrow? The only answer he could arrive at was that a poacher was using bow and arrow to hunt forbidden game. Though the swamp had become a wildlife refuge two years before, and the killing of animals was no longer allowed, it was no secret that illegal hunting went on. But why an arrow modeled after those used by the ancient mound-builders? Perplexed and baffled, Jeb was forced to admit to himself that nothing really made any sense.

That night, Jeb again found it difficult to go to sleep. And it was not nightmares of lost fishermen that caused him to sleep fitfully after he finally dozed off, but dreams of a slender, polished arrow with a finely hewed head and brilliant red tail-feathers.

7

JEB DIDN'T NEED AN ALARM CLOCK TO AWAKEN HIM the next morning. The excitement that had made sleep difficult the night before, now prodded him to early action. Bounding out of bed, he dressed quickly, for the sun had not yet risen and the air was chilly.

Mac, asleep at the foot of the bed, stirred fitfully before jumping straight awake when he heard Jeb's footsteps. Startled at his sudden awakening, Mac started a low growl.

"Sh!" Jeb put his finger on his lips and Mac quieted. Jeb's mother and father were not yet awake, and Jeb was careful not to arouse them. Quietly,

84

he crossed the room over the braided rug that muted his footsteps, on past the cypress chest and highboy; then out into the hall and kitchen, still dark with shadows from the night.

In the kitchen, Jeb scrambled eggs for breakfast and shared them with Mac. Jeb was used to getting his own breakfast, for he was often up and away before it was light. Swamp boys and men, too, come and go in the swamp whenever they choose. They are a restless breed and find the cabin useful mostly for shelter in bad weather and for sleeping.

Jeb had learned from early boyhood to be self-reliant in the swamp. Ma and Pa would never become alarmed or think it strange if Jeb went off to fish by himself in the moonlight, or stayed away from home for several days at a time. Why should they worry? Everything Jeb needed or wanted was there in the swamp—food, water, and a boat to bring him home when he was ready to return.

Jeb collected the few things that always went with him in his boat: his fishing rod; a tin cup; a bucket; a compass and matches, both in a water-proof container; and a bottle of insect repellent. After mentally checking the list, he included a coil of rope and several pieces of cold cornbread, which he put in a paper sack. Then he and Mac quietly left the cabin.

Mac trotted happily along at Jeb's heels towards the dock, stopping only once to sniff companionably at the small figure of Nod, curled up on the bank of the creek. When he reached the boat, the dog slipped into his place in the bow without hesitation. But Jeb stood for a moment on the dock, lost in his own thoughts. Dressed in jeans and a long-sleeved shirt, he was protected against the slight chill of the early morning air, yet he gave a little shiver as he thought of returning to the island. A little shiver not of cold, but of excitement and of lingering fear. It was the fear that made him pause.

A flock of ducks, flying high above his head chattered rudely, awakening Jeb from his musing. Then he got into the punt and both boy and dog were off, down the creek.

Poling through the dim light, Jeb didn't stop to ask himself what he would do if he met the mysterious boatsman, and he refused to let his mind dwell on the fact that there might be real danger at the lake. He could only concentrate on the knowledge that somehow or other he had to find the answers that would bring the swamp back to normal for him. Overriding everything else, was his compelling desire, however irrational, to make the swamp the safe, known haven it had been in the past. This only he could do. No one could do it for him.

The long trip to the lake was uneventful, and again Jeb was pleased at how well-trained Mac had become. Snakes still excited him a little, but a quiet word from Jeb kept him in his place in the bow, where only a slight tremor of his body or a soft whimper revealed his uneasiness.

Once at the lake, there was no indecision on Jeb's part. This time there was no thought of turning back. He guided his little punt swiftly across the clear, open water of the lake, straight for the closest point of the island.

When Jeb reached the steep shoreline, he used the length of rope he had brought along to secure his boat to an overhanging tree branch. Then, with only a quick glance around him, he clambered up the steep incline. Ferns, bushes, and even trees, rooted in the spongy, insecure ground, swayed first in one direction and then another. Even Mac's light weight caused the smaller plants to heave up and down.

"Come, Mac!" Jeb commanded, and the dog hastened to join him on higher, more solid ground.

Jeb had decided to start at one end of the island and walk its entire length, exploring as he went along; when he reached the far end of the island, where he had discovered the arrow, he planned an intensive search.

As Jeb and Mac picked their way through the tangled underbrush beneath the oak and magnolia trees, the song of a thrush filled the air. Blackberry bushes grew in dense patches throughout the forest; and Jeb stopped occasionally to pick a few early berries, while Mac sniffed at unseen animals hidden in the brush.

Jeb was glad he had gotten an early start, for, as

he trudged steadily on, the day began to turn warm and humid. Finally, he reached the interior of the island, where the oaks and magnolias gradually diminished, giving way to a heavy forest of pine and cypress.

Jeb walked for a long while through the cool pine forest, until he suspected he was getting close to the area where he wanted to concentrate his search. He cut over then to the shoreline and followed it until he reached the small cove where the fishermen had beached their boat. It was in this region that he had heard the mysterious, high, twanging sounds; it was also near here that he had found the arrow.

If a bowstring had made those weird vibrating noises, then two arrows must have been shot. One arrow he had at home. The other he would search for.

Calling Mac to heel, Jeb made his way to the spot where the two fishermen had camped. Then after getting his bearings, he was able to find the tree from which he had pulled the arrow. He noticed that only a faint wound remained in the bark of the tree as evidence that an arrow had ever been imbedded there.

Jeb reasoned that the second arrow should be in the same vicinity as the one he had found, for the two sounds had seemed to come from the same direction.

Jeb soon found, however, that looking for the

second arrow was complicated. There were hundreds of places where the arrow might lie hidden. It might be obscured by thick, low-growing bushes, or it might be buried in the sandy soil. If it had been aimed high, it could be embedded in a branch of a tall tree. And there was always the possibility that the person who shot it had picked it up and taken it away.

Still, Jeb hunted doggedly on. The perspiration was running down his cheeks, for it was now hot and muggy. A thunderstorm seemed in the offing. He continuously swatted at sand flies until he remembered to swab some of his insect repellent on his cheeks, forehead, and arms. Though there was no real need for hurrying, Jeb hunted on in a frenzy, driven by his compelling need to solve the mystery of the swamp.

Gradually, he and Mac worked their way deeper into the pine forest. And there, suddenly, in a small clearing, he saw five magnificent wild turkeys, the wariest birds in the swamp. The turkeys froze momentarily, while the light glistened on their iridescent bodies, and then scampered noiselessly for cover in the thick underbrush. As Jeb watched them disappear, the thought crossed his mind that the arrow he had found might have been aimed at one of these birds. If so, the second arrow might have been aimed at one of them as it was making its getaway into the bushes. If so, the arrow would be impossible to find in the impenetrable growth.

Bitterly disappointed at not finding the second arrow, and with his hopes of finding any other clue rapidly diminishing, Jeb wandered on aimlessly. Would he ever find the answers he needed to feel secure again in the Okefenokee? Or would he always have to live with the feeling that some unknown danger lurked somewhere in the hidden depths of the swamp's dark forests?

Somehow, the mystery seemed like an unfathomable jigsaw puzzle. So far, Jeb had put together only the background pieces of island, lake, and forest. And on one side of the puzzle he had filled in two small solid patches. One of an arrow in a tree. And the other, of a boat almost swallowed up in an impenetrable forest. But the most important pieces were missing, for there was no person in the picture and no hint of a central theme or action.

Finally, hot, weary, and discouraged from his long hunt, Jeb made his way out of the forest and over to the edge of the island, where he sank to the ground beneath the shade of a large cypress. From his resting place, he took stock of his surroundings, which seemed somehow vaguely familiar. He was out of the pine barren now and close to the swampy border of the island. First he looked down at the sparkling water of the lake, but the reflected sunlight was so intense he could not gaze there for long. So he let his eyes travel across the narrow stretch of the lake to the cool, green of the forest. For a while, he stared blindly at the thick, green jumble

of bushes, plants and trees, while he wondered what
he should do next.

Then, suddenly, he jumped to his feet so quickly,
he almost knocked Mac down the steep slope of
the bank. For his eyes had focused on a narrow
channel that went behind the curtain of the forest.
Although Jeb had been staring at the same spot for
several minutes, he had almost missed seeing this
strange exit from the lake. It was almost imper-
ceptible, for plants and bushes arched over it just
three feet above the surface of the water. It was like
a miniature tunnel opening up through the dense,

overgrown jungle. This, then, was the hidden boat run through which he had seen the phantom boat disappear.

Jeb began to get an uneasy feeling again, but he shook it off. He could not afford to lose his nerve. Besides, his desire to find the answers to the many questions rushing through his head would not let him give up now.

For several moments Jeb sat and looked at the thick growth of plants that surrounded the hidden channel. It seemed to him that the matted plants were like a screen that divided the swamp into two worlds. The world he knew and a secret and unknown world on the other side.

While Jeb gazed at the forest, he wondered what he should do next. His decision was almost made for him. No matter how frightening it might be to venture into the unknown, he had to do just that. There was no other choice. He had to find the answers to the mysterious happenings in the swamp now, or lose his self-respect forever.

As soon as he had made the decision to go through the hidden channel, Jeb felt curiously relieved. And with his mind made up, he began to pay attention to Mac, who was jumping about, giving short, demanding barks. Mac was hungry. Jeb whistled to the dog and together they trudged the long, hot way back to the boat at the opposite end of the island.

Once in the boat, Jeb brought out the thick

pieces of corn bread, which he had carefully stashed away, and he and Mac wolfed them down. Then Mac leaned over the side of the punt and gulped in great swallows of the cool water, while Jeb quenched his thirst by dipping up lake water in his tin cup and letting it trickle down his parched throat.

By now a sense of adventure had so overcome fear that Jeb was restless and impatient to return to the far end of the island. Eagerly, he poled his boat back to the region where he had seen the hidden channel.

But when he guided his punt along the narrow portion of the lake, and searched the thick, jungle-like growth of the forest for the opening, he could not find it. Disappointed and baffled, he cruised past the forest time after time without success.

Then, he had an idea. Breaking off a small twig from a tree, he tossed it into the water, and watched as the slow current drew it across the lake towards the forest. Carefully, Jeb followed the floating twig in his punt until it finally disappeared behind a thick bush, growing in the water a few feet out from the edge of the forest. Jeb hurriedly rounded the bush, just in time to see the small twig disappearing into the green-tinged depths of the small tunnel that he had been searching for. Now Jeb understood why he had been able to see the hidden channel from the high bank of the island but had not detected the opening from his low-lying boat.

Curious, Jeb peered into the dimly-lit opening

of the run. His heart was beating wildly, and once again he was afraid, but the impulse that had compelled him to return to Emerald Lake drove him relentlessly on. He took a deep breath and, with a stubborn set to his jaw, maneuvered his punt even closer. At the entrance to the tunnel, Jeb hesitated for only a moment. Then, bending low in his boat, he pushed aside the thick plants with one hand, and shoved on his pole with the other.

Quickly and silently his little punt slid through the arched opening. In one bold move, he had entered the hidden channel. He had passed the border from his part of the swamp into an unknown world.

8

AFTER HIS INITIAL PUSH THROUGH THE TANGLE OF vines and bushes at the entrance to the channel, Jeb found the going much easier. Although he had to bend low in his boat to keep from scraping his head against the thick plants that formed an arched roof, there were no stray vines or branches jutting out from the sides to tear or scratch him. It was obvious that the tunnel was used as a passageway often enough to keep it clear of such unwanted growth.

For several minutes, Jeb moved slowly along the strange run. He became acutely aware of the sweet smell of crushed reeds and grasses from beneath his boat, and of the moist, pungent odor of the green bushes that arched over him. He also had time to reflect on his good luck in not finding any snakes draped on the branches of the bushes under which he was passing. For it would have been difficult, if not impossible, to keep moccasins or rattlers from dropping into his slow-moving boat.

At last Jeb could see bright light coming from an opening at the end of the green tunnel, and he felt sure that he would soon exit from the long passageway.

When his boat finally shot out of the thick forest growth surrounding the tunnel, he could only gaze in wonder at the lake that lay before him. A lake that fully equaled the jewel-like beauty of Emerald Lake on the other side of the passage, and that was almost identical in size and shape with it. In fact, so alike were the two lakes, that Jeb had the odd sensation that he was now on the other side of a looking glass. However, there was no large, forested island in the new lake. Instead, there were numerous, small, floating islands, much like the hundreds that Jeb had seen many times in other parts of the Okefenokee. Restless islands that continuously floated hither and yon on ceaseless journeys through the endless swamp waters. Most were large enough to have several trees growing on them, and around their swampy edges there were heavy masses of ferns and water plants. But Jeb had no intention of climbing onto any of them, for such floating masses of roots, covered with leaf-mold and a thin layer of sand, would quake and tremble under his weight and give him no sure footing.

Skirting around one of the islands that floated almost directly in the middle of the large, oval lake, Jeb saw that this lake, too, was completely surrounded by a thick forest. But as he poled around it, he found several channels that formed distinct exits from the lake into the cypress bay beyond.

After seriously surveying the entrances to three of the largest channels, Jeb brought his boat to a standstill while he pondered what he should do next. Motionless, he became aware of how hot and still the air was, so moist and thick it was difficult to breathe. Everything was strangely quiet. Not a leaf stirred on the trees, and his boat barely moved on the glassy water. And it, he now noticed, had changed color from a clear green to a dirty, leaden gray. Looking up, Jeb saw that the water only reflected the masses of gray clouds that were rapidly scudding across the sky and building up into a tremendous thunderhead. Mac, too, had noticed the peculiar change in the weather, and he twitched his nose back and forth, nervously sampling the air, trying to understand it.

Even setting aside the ominous look of the sky, Jeb knew that if he ventured any farther, he would be disobeying every lesson and law of the swamp that his father had ever taught him. There was no doubt in his mind that he should learn the region in and around the new lake before he went deeper into unknown territory. But all his inner drive for a solution to the mystery urged him not to stop at this point.

Knowing that he should not enter a strange forest unless he could recognize a landmark, he did take time to scrutinize the trees. Of the three channels, only one had distinctive markings that would serve to guide him on his return to the lake; at the entrance to the middle channel there were two tall cypress trees with their tops stripped of leaves.

Jeb stopped only long enough to fasten the sight of them in his memory. Then, with his strong impatience to continue the search overriding almost all caution, he made his decision. He would enter the middle channel and cautiously feel his way along. To ease his conscience, he told himself that he would be extra careful. He would go only a short distance, and then return.

The boat run through the heavy growth of cypress was extremely dark. Only a little light penetrated through the canopy of leaves to the water below. In fact, the leaves were so dense that Jeb rarely caught a glimpse of the sky above.

The boat run was not only dark, but very narrow and winding, and Jeb needed all his skill to thread his boat around the protruding cypress roots and knees. Fortunately, no forks or side channels led off anywhere, and he was sure he would find it easy to retrace his passage. Although he had felt a little uneasy at the threatening sky and the unusual mugginess of the day, he had managed to shrug this off. What if it did rain? He had been caught in swamp thundershowers many times in the past. Rain was a common thing, even severe storms.

As the channel went on and on, his main interest became looking ahead for an opening that would indicate he was emerging from the dark bay into a lake or prairie. No such opening came into sight. Jeb finally was impatient at not finding a turning-around point, either at a widening of the run or at its emergence into a larger body of water; but he did not really become disturbed. For he felt he could not possibly lose his way in such an uncomplicated channel. Still, the caution of a whole lifetime in the Okefenokee made him aware that he should go no farther into unknown territory until he had digested this part of the journey. So he made plans to turn around as soon as possible and head back to Emerald Island. After spending the night there, he could go on with his search.

Suddenly, a deep, ominous clap of thunder rumbled through the swamp, reverberating from tree to tree, and a chill wind began to blow ripples on the water's surface. Quickly, the fitful gusts of wind increased until they became a steady force, moaning high among the branches of the tall trees, and then shrieking down to the water far below. A few scattered raindrops began to penetrate the thick, overhanging leaves. Then the few drops became a downpour, so hard and vicious that Jeb and Mac were soon drenched. Just when Jeb decided that he would have to try to pole the punt stern first out of the run, he saw a patch of thick, gray light ahead. The channel was finally emerging into an open stretch of water. There he would be able to turn his punt around and head back the way he had come. One hundred more feet to go, and the boat would be in open water.

As Jeb emerged from the bay, he was hit full force by the ferocious storm. Almost immediately, his punt was whipped away from the protection of the cypress trees, and he found himself afloat and out of control on an expanse of water whose size he could only guess at, for thick sheets of rain obscured his view. The wind whipped buckets full of water against his face, and his punt tossed helplessly about. Every attempt he made to control it availed for nothing. The wind was howling in fury, and the boat was filling dangerously with water. Mac whimpered and Jeb could just barely make out the small, wet figure in front of him. He tried shouting encouragement to Mac, but his words were torn from his lips and lost forever in the wind's wild shrieks.

Jeb tried desperately to turn his boat about so that he could reach the channel from which he had just come; but he was helpless. He could see nothing; and what was worse, the constant battering about of his boat had caused him to loose his bearings. No amount of boatsmanship in such a furious storm, with such a light craft, availed for much. Finally all he had time to do was to frantically bail water from his boat, which filled up almost as soon as he dipped it out. All thoughts of steering were put aside. The only thing that mattered was keeping the little punt afloat. Jeb had no idea where he was. He was in the open water of a lake, but he had no way of telling how large it was. The wind drove relentlessly on. Caught in such a fury of wind and rain, the only chance for survival was to ride out the storm by keeping as much water as possible out of his boat. For an unbearably long time the little punt tossed aimlessly about. Jeb had no idea of how far he had traveled; and there seemed to be no direction to the wind as it buffeted him from all sides, first one way and then the other.

Once, abruptly, the boat scraped over a log or heavy root and almost capsized. But the wind soon whipped it free. A short while later, the punt again collided, this time with a tree. Now, Jeb realized that he was being driven into another cypress forest. Here the darkness was almost absolute. Reaching out his hand, Jeb could feel branches and trees, but he was traveling so fast, he could not catch hold of anything to slow him down. Desperately, he clutched at a limb, but it broke off in his hands and almost threw him out of the boat.

The storm was now so fierce that the cypress trees were no protection from the fury of the wind. The boat smashed against another cypress root and then another until Jeb wondered how much punishment the frail craft could take. At each impact, the shock was so great that Jeb was sure both he and Mac would be thrown out. But the dog was still with him, although he had deserted his place in the bow of the boat, and had slowly inched his way along the bottom of the punt until he felt the comfort and warmth of Jeb's legs. He lay quietly, occasionally reassuring himself by licking Jeb's hand.

Jeb was close to fatigue and could only hope and pray that the storm would soon subside. He bailed blindly and steadily until his arm muscles and shoulders ached. The boat constantly smashed against trees in the cypress bay, and then was sent

helter-skelter deeper and deeper into the unknown forest. Even Jeb's boat pole went, washed overboard after a vicious collision with a large treeroot.

Jeb had no idea how long the terrible ordeal lasted. In almost pitch darkness, with all the strength and fury of wind and water pitted against him, it seemed to go on forever.

Even when the wind did abate somewhat, there was not much Jeb could do but ride along with his boat and hope for a place where he could finally anchor it. From blackness, Jeb found himself moving into a leaden gray cloud of mist that blocked his visibility as much as the total darkness had. Although he still could not see where he was going, the change in lighting and the fact that he was no longer being shoved into roots and trees, made Jeb realize that he was probably clear of the forest. Still, without his boat pole, he was helpless.

Then Jeb felt an impact and the boat stopped. Though turbulence still rocked it back and forth, the craft had somehow become anchored.

Gingerly, Jeb reached an arm over the side, groping blindly in the thick, soupy mist. He shoved his arm elbow-deep into the water until his fingers touched the bottom. With a surge of hope, he realized that he was in shallow water. And the soil he touched was wet and mucky, the kind of soil that surrounds both large and small islands in the swamp.

9

CAUTIOUSLY, JEB PUT ONE LEG OVER THE SIDE OF his boat and immediately sank knee-deep in the soft mud. Then, carefully holding on to the side of his boat, he swung his other leg over into the water. Moving slowly, he took a number of steps away from the punt, and found that the soft bottom gradually became firmer. Returning to the punt, he pushed and shoved until he finally dislodged it from the sand bar upon which it had become jammed. With only Mac in the boat, he was able to drag it after him with comparative ease. As he walked on, the earth became less and less mushy, although it still quivered beneath him with each step. Slowly, he continued on to what he hoped would be firmer ground.

Finally, the water became so shallow that it only came up to his ankles. At that point, Jeb ordered Mac to jump out and, a short distance farther on, he pulled the boat out of the water onto a spongy shore. Stumblingly, he pushed and pulled it over the slippery surface until, exhausted, he sank down beside it on ground that was wet but satisfyingly solid. At last, he had reached the stable soil of an island.

Jeb was tremendously relieved to be on solid ground again, and he marveled at his good fortune in coming through the storm with his boat still intact. The wind had ceased its terrible force, but it was still unpleasant and cold, and strong enough to whip his wet clothes about him, chilling him through and through. But still he was lucky.

With hesitation, fearful at what he was sure he would find, Jeb began to fumble through the interior of the boat. Most of his possessions had been lodged tightly in the bow, but there seemed little chance that any of them would be left after the terrific pummeling the boat had taken. He knew that his boat pole was gone, and he suspected that the bucket he had used to bail water had long ago floated over the side. But miraculously, the bucket, half full of water, was still in the punt. He also found his tin cup; waterproof case, complete with compass and matches; and fishing rod still in the bow. As tired out and soaking wet as Jeb was, things looked better just knowing that his valuable possessions were still intact.

The rain set in again, in a dreary downpour. But the wind, though blowing steadily, now came from one direction only, enabling Jeb to lean his boat against a tree as a windbreak. Then, completely exhausted, boy and dog curled up on the wet ground, huddled against each other for warmth, and slept until the dawn awakened them.

Dawn was a bitter disappointment to Jeb, for, although the rain had stopped, heavy mists still billowed and rolled about, making visibility extremely poor. Jeb was stiff from his night on the wet ground, and he desperately needed a fire to take some of the cold and wet from his body. He was positive that he would lose his sense of direction if he wandered very far from his boat, so he ventured only a few feet from the windbreak to grope for wood.

Surprisingly enough, he got some damp leaves to smolder and then a few pieces of wood caught fire. The heat from the feeble blaze wasn't much, but the sight of the small flame cheered him immeasurably. Jeb huddled for some time over the small fire, occasionally getting more wood to revive it. Mac, too, moved close to the faint warmth which it gave out and, with a great sigh, accepted the damp, cold world.

During the time that Jeb had concentrated so hard on surviving, his mind had been as numb as his body. But now that the danger from the storm was over, he had to face up to his predicament. With a sick feeling in the pit of his stomach, he had to admit to himself that he was lost. Lost! He, Jeb, a swamp boy. And it was his own fault! He had followed the long channel when he knew better. He had misjudged the ferocity of the storm. He had done everything wrong. And it was mostly because he had been running away from his own fear. Well, there was plenty to be afraid of now.

For several hours, while mist and fog swirled about him, keeping him bound close to his boat and small fire, Jeb had plenty of time to curse his foolhardiness. Glumly, he realized that he had even muffed his chance to find a solution to the mystery. Because of his misguided zeal, he was farther than he had ever been from finding the truth.

Eventually, when the mist cleared somewhat over the island, Jeb decided he must search for food. Thick vapors still swirled up from the surface of the lake and he realized it would be many hours before he could see to fish there. So he made his way along the shoreline, while Mac followed behind him like a small, forlorn, black shadow. Jeb walked about a mile before he reached the other end of the island. Then, he decided to trek back to his boat by a different route, so he cut over to the interior, searching for berry bushes along the way.

In the central portion of the island, he found a narrow stream, which seemed to cut the island into two long halves. He followed the stream for some distance, and realized that he had made a mistake in leaving his fishing gear in the boat, for there were many fish darting about in the clear water. Intrigued by the sight of so many fish, Jeb kept watching the stream as he hurried along. When an exceptionally large pike swam into view, Jeb watched, fascinated, at the light reflecting from its silvery scales.

Then it stopped. For some reason, it seemed unable to swim away from a small area. It appeared trapped.

Jeb fell to his knees on the bank to examine the fish more closely and was startled to see that the fish was a prisoner in a small, carefully made, wooden fish-trap. In stunned disbelief, he noticed every detail of the trap. It was cleverly shaped and contrived, so that a fish could enter from one end but could not then escape because of many, small, wooden pegs. Carefully, Jeb lifted the trap from the shallow water and removed the large fish. Then he turned his attention back to the trap, marveling once more at the skillful way it had been fashioned.

It was not only the workmanship, however, that was fascinating. It was also the fact that the trap existed at all; this could only mean that he was not the only person who knew about the island. In wild excitement, Jeb jumped to his feet and eagerly scanned the surrounding forest. But there was no sign that anyone came here often. Whoever had set the trap obviously did not live here, and might not be back for a long time. He was still lost.

As he examined the trap again, it occurred to Jeb that it might be another clue. Another piece to the picture puzzle he had been slowly building up, bit by bit. Maybe this trap was only one of many, set in a number of suitable places, and connected with the arrow and the strange boat. Just the thought sent a shiver of apprehension up Jeb's spine, for he still did not know whether the phantom boatsman was a person to be feared or trusted.

But whatever the answer might be, Jeb felt a surge of gratitude toward the unknown person who had unwittingly provided him with a meal. Then, knowing that other fish would find their way into the trap, Jeb replaced it in the stream so that its owner would not find it empty when he came back to it.

When Jeb returned to his boat, he hastened to prepare the fish for roasting. A quick, hot fire, a green twig for a spit, and soon the wonderful aroma of cooking fish was driving Mac into happy fits of barking. Never had a meal tasted so good. The fish was soon eaten and all the bones picked clean. Then, since both he and Mac were still hungry, Jeb went back to the stream with his fishing equipment and caught more pike.

After another feast of fish, this time enough to quiet the hunger of both Mac and himself, Jeb prepared to bed down for the night. First, he gathered enough wood to keep a roaring fire going. Then, full and warm, he gradually began to lose the tight feeling of panic that had been with him ever since he had become really aware that he was lost.

10

THE NEXT MORNING, JEB AWAKENED TO A HUMID, overcast day, still bleak with storm clouds. A weak sun, however, showed through breaks in the cloud cover, and Jeb's hopes rose as he struggled to rise. His clothes were dry now, and Mac's fur was once more soft and fluffy. During the night, the fog had almost disappeared from over the water, and visibility was good enough to allow Jeb to examine the large lake that surrounded the island. He noticed that the lake seemed to merge with vast stretches of water meadows, which were broken by small ponds and numerous channels. In the far distance, on all sides, were thick, dark cypress bays.

It was from one of those dark bays that Jeb had been driven during the terrible storm. And one long glance told him it was useless to even try to guess which one. Looking over the lake, he felt again a sharp touch of panic, and of regret at his own foolishness.

He bit his lip as he looked out over the strange marshes, lake, and forest. Away it stretched, an alien swampland. On and on, as far as the eye could reach. And in all that space, it seemed that he and his small black puppy were the only living creatures. Angrily he tried to shake himself out of his mood. This was no time to give up. He might be lost, but he was not helpless.

Jeb knew that he had to find a way back to familiar territory, but there were two things he had to attend to first. Breakfast, and finding a new pole for his boat.

Once more, Jeb was thankful for the bounty of fish in the Okefenokee, for an hour's fishing rewarded him with more than enough breakfast.

After a quick meal of roasted fish, Jeb headed back toward the interior of the island in search of a long straight pole that he could use for punting. Luckily, the storm had ripped many branches and twigs from the trees, and Jeb soon found a tree limb that was suitable.

Back at his camp site, Jeb removed his compass from its waterproof case and stowed the rest of his equipment in the boat. Then he and Mac shoved off from the island. Completely lost, his only hope now for getting out of this wild region and back to the world he knew was to head constantly in one direction. So, a few feet out from shore, he paused to get his bearings from his compass. It was only logical that he try to work his way westward, towards his home on the flat pinelands that edged that boundary of the great swamp. So, checking his compass once more, he steered towards the west and the dark cypress forest in the distance.

He had thought briefly of awaiting the return of the person who had made the trap. But there was no guarantee of when that would be. So, though he knew well what a long and tortuous trip it would be, through the perpetual labyrinths of the sunless bays, there seemed to be no other choice. It might take him days or even weeks to get back to familiar territory. Even with a compass to guide him, he knew how difficult, how impossible, it would be to continue long in one direction on those twisting waterways. If only he could go in a straight course westward, like the sun in its journey across the sky. But he was bound to his boat and would have to twist and turn with the channels, unpredictable channels that would force him first in one direction and then in another, compelling him to go north and south and east, when he wanted only to go west.

By the time he reached the forest, the sun was shining brightly, and only a few, white clouds still marred a clear, blue sky. After carefully surveying various channels, Jeb finally chose one with a broad passageway that seemed to go directly west and plunged ahead.

The air in the bay was cool and rich with the fresh fragrance of the recent rain. Drifting along, Jeb looked up at the formless masses of Spanish moss, clinging to the thick foliage of the tall trees, and he thought that they looked like fragments torn from the gray morning mist of the swampland. There were multitudes of snakes draped on bushes and cypress roots. Jeb could see rattlers and cotton-mouths coiled up together in groups, but they seemed in some sort of stupor after the storm and neither noticed him nor moved as his punt glided by.

On and on he drifted, under the towering trees. Again Jeb felt that he was in an upside-down world, for when he glanced at the water he saw the clear reflections of countless trees, occasional clouds, and once the curious masked face of a little raccoon. On and on he went, through the quiet, dream-like setting of the dark, cypress bay.

As Jeb continued his silent journey through the deep forest, he found time to reflect clearly on his position. Although he was certainly lost, he had his compass and, while he was searching for a way out of this wilderness, he knew how to live off the swamp. In the swamp, there was good water, plenty of food, and Jeb had his precious matches, which meant he could have fire for warmth and cooking.

What he felt now was self-contempt. He had thought he knew the swamp too well ever to get lost in it. But he was lost and his main job was to find his way out by himself, without the humiliation of having a search party come to find him. He had stayed in the swamp before for several days at a time, and so felt fairly certain his father would not become concerned unless more than a week went by. Unless, of course, the storm made him worry. That was possible.

Jeb also had time to reflect on the events that had led him into this unchartered region in the first place. Again, he recalled the high, vibrating sounds; next the phantom boat; and then the Indian arrow. And, of course, there was the more recent discovery of the mysterious and well-made fish-trap.

Once more, Jeb found himself peering off into the surrounding shadowy bay, his eyes searching to see if anyone was around; for certainly the arrow and fish-trap were evidence that someone else knew this region. But there was no sign of another human being. Jeb was completely alone. He reached forward and ran his fingers through the thick ruff on Mac's neck. In the loneliness of the great swamp, Jeb was truly grateful that he had Mac along.

"Good old Mac, we'll get out of the swamp before long," Jeb said. Recently, he had found himself talking aloud more and more to his dog, who responded with an understanding look in his dark eyes and a series of happy thumps as his tail hit the wooden floor of the boat.

For long, weary hours, Jeb wandered on through unknown prairies and dark runs in the gloomy cypress bays, trying to keep a westward course. Still he encountered no familiar scene. Disappointed, he kept poling along with nothing to go on but hope. Finally, hot and sweaty, he stopped and gathered a few leaves from a large bush called "poor man's soap." Crushing the leaves in his hands, he dipped them into the water, rubbed vigorously to produce a frothy lather, and then washed his face and hands. Next, he removed his shirt and let the cool water slosh over his hot body. Refreshed, once more he started boating.

Presently, Jeb realized that the sun was slipping lower and lower in the sky. Soon he would have to find an island where he could spend the night; but as he stared out over the prairie landscape, he saw nothing but endless marshes and small, floating islands. These floating mats of earth and vegetation were like giant sponges, and were so sensitive to any weight that he did not want to chance stepping onto any of them. So he poled steadily on until, in late evening, he saw a dark island in the far distance.

Even from a distance, there was something foreboding about the island. But gloomy as it looked, Jeb knew he would have to stop there for the night, for the sun would set before he could find anywhere else.

From the minute that Jeb and Mac stepped onto

the island, the going was rough. Live oak, pine, black gum, cypress, and holly trees grew close to the water, and saw palmetto crept beneath the trees to form a thick, jungle-like growth. Thorny thickets prospered in the soft earth, and bordering the thickets were dense masses of ferns and maiden cane. Jeb, whose hands were filled with his fishing gear, bucket, and matches, elbowed his way through the waist-high growth to make a pathway for Mac, who followed close on his heels.

Even when they were past the thickets, the route was almost as difficult. Heavy trees were covered with vines and climbing plants so thick they looked as though they were choking the trees. Nevertheless, they continued on until they came to some pine barrens.

Several large ponds dotted this region, and Jeb carefully skirted around them, keeping well back from their borders, for he suspected that they harbored alligators.

Although Jeb saw no sign of an alligator, he did notice a huge mound of sticks, twigs, and leaves piled up in an opening between the trees. He knew that near the bottom of this large pile of decaying vegetation were alligator eggs, which heat from the sun and the rotting debris would incubate. Jeb rushed to the mound and started searching near the surface, not for alligator eggs, but for terrapin eggs, which that lazy animal often lays in the topmost layer of such piles of trash. After digging through several inches of the trash, Jeb found what he was hunting for: a large pile of elongated, white terrapin eggs. He was delighted with his find, for terrapin eggs are a great delicacy and neither he nor Mac had eaten since breakfast. Carefully, he took off his shirt and tied the sleeves together to make an impromptu cloth sack to hold the eggs. Then, he took them to a spot where he could build a fire, and carefully laid his bundle down on the ground.

Mac trotted along as Jeb searched for firewood, and later rested on his haunches while Jeb arranged the kindling for a fire. With thoughts of hot food to occupy him, Jeb was almost able to shake off the gloom of the island. Carefully, he broke half of the eggs into his bucket, which he planned to use as a cooking vessel. The rest, he decided to save for breakfast in the morning. A green twig made a perfect mixing spoon, and Jeb was soon beating the eggs into a light, frothy mixture.

The eggs ready, he struck a match and shielded the small flame with his hands while he set fire to a tiny pile of wood-shavings. He was so busy trying to fan the feeble flame into a larger blaze that he didn't see or hear the terrified deer until she had broken into the clearing and was almost upon him. As Jeb looked up, startled by the sudden noise, he saw a little doe who had obviously been so badly frightened by something in the woods that she was oblivious to everything but her pursuer. In her fear, she skirted so close to Mac and Jeb, that Jeb could actually see the terror in her wide, brown eyes before she went crashing away into a thicket.

Immediately, Jeb became alert, for he wondered what had frightened the doe. He didn't have long to wait to find out, for following close after the deer came one of the most dreaded animals of the great swamp—a wild boar. It was an old boar, huge, gray and ugly. His small, vicious eyes gleamed red in the half-gloom of the forest as he searched in the dim light for the doe. Jeb signaled with his hand for Mac to stay motionless. Then, both Jeb and Mac watched anxiously while the savage beast, with its poor eyesight, rushed by, without seeing them.

Jeb felt weak, and he wiped beads of perspiration from his forehead; for the boar had caught him entirely off guard. He knew that he would spend an uneasy night; with a carnivorous hog roaming the island, neither he nor Mac was safe.

Hastily, Jeb smothered his fire and moved his belongings away from the cooking spot, for the area might be part of a path that the boar used regularly. Then, working quickly, he carried branches and logs to form a semicircle in front of a low-branching pine tree. He and Mac would sleep with their backs against the tree and with a wall of fire in front of them. The fire might ward off an attack by the hog, if he should return during the night. Not until they were settled behind their barricade of fire, did Jeb turn his attention again to food. But with the added worry of the boar, the edge had somehow been taken off his appetite.

For a long time, Jeb watched the fire leap into the dark sky, while showers of crimson sparks were whipped about by the wind. The night was very, very dark, for the moon had not yet risen. In fact, Jeb could see little beyond the small area lighted by the fire.

He thought back now, with compassion, on the terrified doe and hoped silently that she had kept her lead. If so, with the island so dark and black, she might still be safe. For it was the kind of night that favored all hunted things.

Jeb leaned back, exhausted, against the trunk of the pine tree and closed his eyes, but anxiety over the boar kept sleep away. Each time he shut his eyes, an image formed of the warm kitchen at home, with Ma's face smiling as she bustled about the cheery room, rich with the odors of chicken and spoon bread. Unconsciously, Jeb heaved a sigh, as he realized how far off all that was now. He even wondered if he would ever see her again.

Finally, his weary mind stopped remembering and wondering, and he fell into a troubled sleep. But he napped for only a short period before he awakened to keep another anxious watch.

Noises of the forest helped him to stay awake for short periods, during which he fed the fire to keep it from going out. After each watch, he slipped into another short, shallow sleep, only to be awakened again by an owl or some other wild sound from the night. Often, it was a big, bull gator's bellowing voice, roaring for miles across the swamp, carrying with it a musky odor. Sometimes, it was a scream from a wildcat, or the bark of a fox.

Besides building the fire for protection, Jeb also had plans to escape upwards, if need be, through the low-growing branches of the pine tree that sheltered them. In case of an attack, he would try to climb it and hoped fervently that he would be able to carry Mac up with him.

Jeb awoke abruptly at dawn. The light overhead was a pale gray color that seeped through the trees, giving an eerie appearance to the pine woods. At first he could neither see nor hear anything that might have awakened him. Even the old owl had ceased his hooting. The fire was only a pile of smoldering embers, and Jeb knew that he should get more wood. But he hated to make any noise that might draw the boar's attention to him if it should be nearby. Uneasily, he sat with his back against the tree and Mac's warm head upon his lap, and listened.

Then he heard the sound that had probably awakened him. It was a rustling and scraping made by the boar's hard hooves as he dug up the roots of a young pine tree for a meal. The noise got louder and louder as the animal rumaged closer and closer to where Jeb and Mac were camped.

And then with a crash the ugly creature emerged from the brush, only about twenty feet away from the pine tree. Now, Jeb had to make a quick decision. Should he stay quietly where he was and hope that the boar wouldn't see him, or should he attempt to climb the tree, carrying the dead weight of the dog?

Jeb decided to get out of the reach of the wild animal by climbing the tree. Mac moaned softly in his sleep as Jeb carried him in one arm and reached up for his first hold on a low branch of the pine tree.

But it was too late! Even before Jeb could shift his weight and put one foot on the first branch, the

boar began to charge. Jeb was trapped! Unbelievably swift, the old hog came on, his wicked eyes gleaming and his four curved tusks looking as sharp as daggers.

Mac, who had been asleep in Jeb's arms, awoke with a low growl. The fur stood up in a long, stiff line down the middle of his spine. His ears were drawn back, and his teeth were bared. A low growl escaped from deep down in his throat.

With a bound, he was out of Jeb's arms and

charging straight at the wild boar. Jeb stood and watched sick with fear, for he knew that even grown dogs, trained to fight and kill the vicious hogs, were often gored. But Mac was young and quick. He brushed his teeth against the thick side skin of the heavy beast, and then jumped nimbly to one side.

The boar completely forgot Jeb, as it turned its attention towards Mac.

Again and again, the quick and agile dog darted in towards the snarling beast, and then turned quickly away, out of reach of the sharp tusks.

Jeb did not dare call to Mac, for any wavering of the dog's attention, even for a moment, might prove fatal. Still, the fight was so uneven, that Jeb shuddered to think of the outcome. It was a strange dance that could surely lead only to exhaustion for Mac and victory for the ruthless, killer hog.

Time after time, Mac darted in almost within reach of the gleaming curved tusks of the boar. The old boar had plenty of space behind him and stepped backwards as the dog charged, then made vicious stabs as Mac came within his range. Slowly, the hog edged backwards, and all the time his small, wicked eyes never left Mac. He was cunningly waiting until Mac misjudged the distance that his razor-sharp tusks could reach. Then he would put a gruesome end to the contest. Mac's frantic barking never stopped. Over and over again, he moved in towards the boar, who always stepped carefully backwards before slashing out ruthlessly at the brave dog.

Suddenly, Jeb realized the strategy behind Mac's relentless harassment of the boar. He was deliberately backing the savage animal toward one of the large ponds. Each time the boar stepped backwards, after a fierce, barking attack by Mac, he came closer to the edge of the murky pond. Some old hunting instinct must be at work.

Jeb held his breath. Maybe Mac could do it!

The early morning light, filtering through the trees, focused on the thick, gray, bristly body of the boar. His massive head was alert and ready for the fatal strike. His small, mean eyes gleamed with hatred. Again Mac charged, again and again. The old boar was now three feet from the pond's edge, now two, and still Mac backed him on, till the boar stood only a few inches from the green water of the stagnant pool.

Mac rushed in for another attack. The infuriated boar lunged at him recklessly, and a tusk caught Mac in the muscle of a foreleg.

Too late, Mac leaped away. The boar had made a strike, and one of its ugly tusks gleamed scarlet with blood. The gallant dog was badly hurt, but he made ready for another attack. Jeb's anxiety increased for now the boar could more easily catch the crippled dog off balance.

But before Mac could charge again, a massive bulk loomed in the pond behind the boar. A huge alligator had surfaced, evidently attracted by the sounds of the fight. Jeb watched, spellbound, as the huge creature flexed its tremendous tail. Then that tail lashed out towards the old boar. The boar turned when he heard the noise, but he was not fast enough to escape the crushing blow, which swept him into the water with a mighty splash. For a while, the water boiled as the alligator disposed of his prey in his efficient jaws.

Jeb was sick at the violence of the scene, but relieved at the unexpected help for Mac. Quickly, he rushed to his injured dog, now weaving uncertainly, weak from loss of blood. Jeb scooped up Mac in his arms and carried him tenderly back to the fire, where he ripped off a strip from the bottom of his shirt to bind the badly bleeding leg. The leg had a deep gash in it and Jeb applied pressure over the wound until the bleeding finally stopped.

Jeb could see the exhaustion in the sick dog's eyes, and he crooned over and over in a sing-song voice, "You'll be all right Mac, you'll be all right."

At last the weary dog closed his eyes and, for many hours, Jeb kept close watch while Mac slept on and on, completely spent from his fierce battle and loss of blood.

11

JEB SAT CRADLING MAC'S HEAD IN HIS LAP FOR A long, long time. The dog now slept fitfully, but Jeb kept up his vigil until Mac really awoke and thumped his tail on the ground apologetically, as though ashamed to be sick and injured. Jeb felt the dog's soft nose and found it dry and hot. His injured leg was now swollen and red, and when he tried to get up he was unable to do more than half lift himself from the ground. Jeb frowned. They were lost and away from any human being, yet somehow he had to get help for Mac. More than ever before, Jeb felt the terrible isolation and loneliness of the great swamp. Even the humiliation of being found by a search party would be welcome, if it meant saving Mac's life.

First though, he wanted to see if he could get Mac to eat. Although the remaining terrapin eggs had been crushed and trampled during the boar's surprise attack, the fish in the swamp waters provided ample food. Jeb made a stew of fish and water. Then, with great effort, he coaxed the dog to eat a little. Jeb ate a little of the stew, too, but he was so upset and worried that the food almost stuck in his throat.

Finally, with great tenderness, he carried Mac to the boat and carefully laid him down. After another ration of water for the panting dog, Jeb shoved off from the dark, forbidding shores. As the punt moved away from the island, Jeb hoped fervently that he would never set foot there again.

Urgently, he poled his boat through the endless swampland, praying all the while that he would soon find a way out of the unknown region. Over and over again he told himself that Mac would not be wounded and ill, but for his foolish actions.

Jeb's determination to get help for Mac drove him on and on, from one prairie to another and through countless cypress forests with their green-tinged canals. Still, he found nothing familiar and no sign that he was any closer to finding his way back to the world he knew. He did not even see a likely island to land on. Each run, each prairie, each thick cypress bay looked so much like every other, that he concentrated only on going west. But although he was able to go west for short distances in a few channels, he lost ground in countless others that twisted him about in unwanted directions.

At last, terribly, terribly tired, he asked himself if he were really making any headway. For it seemed that every time he entered a different bay or prairie, or run, he was going deeper and deeper into a tremendous maze. A maze from which he felt he might never extricate himself. But he had to keep going. He couldn't help himself. Despite the endless setbacks, he had to continue trying to steer in one direction, for that was the only chance he had to work himself into a part of the swamp that he could recognize.

Jeb stopped every once in a while to encourage Mac by stroking his head, or by giving him some cool water to drink. Though Jeb offered Mac more of the stew, the dog was too sick to eat. He only wagged his tail a little before falling into a feverish sleep.

Jeb lost all track of time. His hands became blistered and cracked, and his face burned from the brutal sun. He constantly checked his compass or strained his eyes for a landmark that he knew, even though there was little chance that he would recognize any part of this wild segment of the Okefenokee.

A flock of geese, flying high in a wide V, crossed the vast expanse of the sky. Jeb noted that there was no hesitation in their purpose. No doubt as to their destination as they flew steadily on. They were going home. As their wild, strange cries filled the swamp air, Jeb was overwhelmed by a longing so strong he could hardly bear it. A longing to be like the wild geese. To be going somewhere definite, on a straight course. To be going home!

Afternoon and evening blended into one long nightmare as Jeb searched desperately for a way out of the wilderness. Even after the sun had set, and with only the faint, pale light of the day's afterglow lighting his way, Jeb continued on his restless journey. On and on he went.

Even after darkness had fallen, Jeb continued, blindly choosing the runs he felt would take him west. Although the moon was not yet up, he could distinguish the shapes of trees and bushes as their heavy masses loomed against the darker background of the night. Somewhere a loon wailed its lonesome cry, while from everywhere there arose the calls and cries of swamp animals who were hunting or being hunted.

Never before had Jeb poled in such a frenzy, for so long a time, and eventually he dropped wearily to the floor of the boat and let the slight current carry him along. Occassionally, he dozed off, sitting upright in the punt, his arms clasped around his legs for warmth.

In the middle of the night, he awoke and looked

around him, amazed at the strange beauty everywhere. It had been very, very dark when he had last dozed off, but now a silver moon had risen and wrought a startling change. A deep sense of mystery and silver magic had stolen over the swamp. Silver, silver. Everywhere there was silver. Even the clouds near the moon reflected its silver sheen. For a long time, Jeb looked up through the silver branches of the trees and watched the moon riding high in the sky like a great, pale, silver bubble against the black backdrop of the heavens. All about him, the soft, black, velvet water lapped gently to and fro, trapping the silver magic on the crest of every wave, where it shimmered like great dollops of silver frosting.

Silver, silver everywhere. Jeb and Mac rode on and on, a silver boy and a silver dog in a silver boat.

Only complete exhaustion made Jeb finally curl up on the bottom of the boat in a sleep so deep that he knew nothing more.

The night passed and the early morning found Jeb and Mac still adrift, aimlessly and without direction, in a wide and beautiful lake. And in the center of the lake was a lovely island, surrounded by a dazzling, white, sandy beach. Jeb, half awake, looked at it, pleased. It was the first island of any size he had seen since the boar island. And it was so different. But he was too tired to find his way there. He dropped back to sleep.

So the little boat bobbed on, still without guidance, until by chance it beached itself on the glistening sand of the island. There Jeb and Mac continued their long sleep until the bright light of the midmorning sun awakened them.

12

JEB WAS DELIGHTED, ON AWAKENING FINALLY, TO see the fresh beauty of his surroundings, so different from the dark island of the wild boar. Happily, he stepped out onto the clean, white, sandy beach and stretched his cramped legs.

Mac thumped his tail when he heard Jeb moving about and Jeb, hurrying to him, saw that although the injured leg was still badly swollen, the dog seemed less feverish and a little stronger. A light breeze was blowing, but the sun was hot, and Jeb decided that Mac would be cooler and more comfortable resting under the tall trees that stood beyond the wide beach. So after first giving Mac some water from his cupped hands, he carried him to a green, mossy bed under the trees, out of the glare of the burning sun.

For a while, Jeb sat beside Mac on the soft moss and looked around him at a beauty almost beyond description. Crystal-pure water flicked along in almost geometric patterns under a brilliant blue sky laced with a few dazzling white clouds. As he rested, a large crane skimmed gracefully over the top of the shining ripples before settling down in the shallow water, where it waded slowly in and out of patches of aquatic plants searching for food.

Watching the crane reminded Jeb of his own empty stomach. So after speaking a few words of comfort to Mac, Jeb left to scout the island in search of food for the morning meal.

From the first, Jeb had been attracted by the island—with its sparkling beach, open trees, and brightly-colored flowers. And now, walking along on the white sand, he felt strangely lighthearted. Even the weather was perfect, for although it was hot in the sun, it was cool as he trekked along under trees that bordered the beach. He began to circle the island, being careful to stay close to the shoreline and always within sight of the waters of the lake.

He had not gone far, when he noticed a trail leading through the bushes and trees towards the interior of the island. Deer often make such a trail when traveling from secluded thickets to a favorite drinking spot beside a lake, and Jeb gave it only an idle glance. Certainly, he was not interested in following it for the sake of curiosity. His only purpose was to find food, return as soon as possible to Mac, and then move swiftly away from the island to continue his search for a way back home.

He was just stepping over the trail, when his sharp eyes spotted a clear impression in the sandy soil. For a second, he simply stood looking down at it. Then he fell to his knees to examine it more closely. For it was a human footprint! With great excitement, Jeb examined the sharp print from all angles. But there was no doubt in his mind. Here where the earth was still damp from the morning dew, there was an unmistakable outline of a human footprint. But Jeb was also puzzled. For the print was not of a bare foot, nor yet that of a hard shoe. It seemed, rather, to have been made by some peculiar type of softer footwear.

There was just the one tantalizing footprint, for a little way toward the interior of the island lay a soft carpet of plants, and a short distance in the other direction was the dry sand of the beach. In neither of these areas would an imprint last for long.

Jeb stood up and looked slowly around him, half expecting to see the owner of the footprint. But there was no one to be seen. All Jeb saw were green plants on one side of him, and the white sand, leading to the lake, on the other side.

Jeb felt like an intruder now. For someone unknown to him probably fished and hunted and carried on the daily business of living on this beautiful island. Jeb was struck by the similarity of the situation in which he now found himself, and the one he had encountered a few days before. There, in the familiar swamp, he had heard unfamiliar vibrating sounds and seen the strange boat of an intruder; now he, perhaps, was the intruder.

Could it be that arrow, boat, fish-trap, and footprint belonged to the same mysterious person? And that that person was nearby? It seemed both likely and unlikely.

Jeb felt as if he were standing on a knife-edge of decision. He could beat a hasty retreat back to Mac and his boat and head blindly onward, hoping to fight his way out of this strange region, or he could follow the path and perhaps get answers to the questions that had haunted him for so long. The possibilities slowly revolved in his mind as he pondered what to do.

He longed to follow the path. His drive to solve the mystery of the arrow and the boatsman had never disappeared; only his great fear for Mac's safety and his chagrin at his own foolhardiness had made him abandon the quest. But was it foolhardy now to take this path? What waited at the end of it?

Jeb retraced his steps back to Mac. The dog was still sleeping and his steady, even breathing showed that he was on the way to recovery. His soft, velvety nose was a trifle less hot, indicating that the fever

had broken sometime during the morning. He was
sleeping a refreshing sleep that could only lead to
greater strength. Relieved, Jeb hesitated no longer.
He owed it to himself and to Mac to see what lay
on that path. There might be help there, and he
could leave Mac long enough to find it. He headed
straight for the path. When he reached the spot where
the strange footprint lay, Jeb pressed his own foot
into the moist sand beside it. Then he compared the
two prints for size. Jeb's was smaller in length by
half an inch, but the width of the two prints was
about equal. Whoever had made the print was probably a boy like himself. A boy not much older than
he, who knew the island well, and who would probably be as startled to see Jeb, as Jeb had been to find
evidence that someone strange had been on Emerald
Island.

Jeb started resolutely down the cool, shaded
path through the tall trees. Bright golden sunshine
filtered through the open, lacy branches and made a
dappled pattern in front of him. Birds were singing
from trees and bushes that grew alongside the well-
defined trail, and occasionally, a shy deer or a small
animal scurried in front of him before seeking
cover. But Jeb did not let himself loiter to look at
his surroundings.

For a long time the way was straight and the
ground smooth as he sped along. Then the terrain
started a slow rise, and after a while Jeb found himself on a high hill looking down into a sheltered

valley. And there, nestled within the valley, lay a small village. A feeling of wonder and then of panic seized him. He had not expected so much. He pulled himself into the shadow of a tree and stood motionless, almost forgetting to breathe, as he realized the full significance of the sight before him. For without any doubt he was gazing down at an Indian village.

There were about twenty beautifully made, grass-mat houses, and behind the cluster of homes was a garden. A short distance behind the village proper, the terrain sloped gently upward, and Jeb saw that the earth had been carved into four handsome terraces. At the topmost terrace, the ground leveled off to form a wide, open area. And in the center of this flat region was the most astonishing feature of the whole village—a large, round building. From the ground up, about one-third of the structure appeared to be a gigantic, circular mound of earth, from this rose walls of wood, capped by a graceful, arched roof. In front of the building was a large, oblong courtyard surrounded by a high bank or terrace. And in the middle of the courtyard, a tall wooden pillar, or obelisk, stretched forty feet into the sky, tapering to a point at its upper end.

Jeb studied every detail of the building, courtyard, and obelisk before turning his attention back to the rest of the village. Suddenly, out of the corner of one eye, he noticed a movement. Turning his head in its direction, he saw a thin wisp of smoke rising above one of the houses. The village was inhabited.

He looked about him uneasily, for he knew that he was in a precarious position. Instead of finding one person, he had stumbled upon an entire village. He had come alone, unarmed, as an intruder on an island whose people and village had remained hidden from the outside world until this moment.

But he did not have long to think about his vulnerable state. A sudden, rustling sound caused him to spin around quickly, and in amazement he found himself face to face with a tall Indian boy. There was a scowl on the boy's face, anger flared in his eyes, and in his hands was a drawn bow, the arrow aimed directly at Jeb.

Jeb's thoughts whirled in his head. He could not move; he could only stand there, staring in fascination at the tall Indian lad and the arrow. An arrow whose every detail Jeb quickly recognized, for it was similar to the one that had been discovered in the mound-builder's grave, and identical to the one he had found on Emerald Island. The Indian boy could only belong to the long-vanished tribe of "tall ones," who for centuries had remained hidden, deep in the heart of the great swamp.

Finally, Jeb spoke, "Hello, my name is Jeb and I want to be your friend. I'm lost in the swamp and need your help."

Still, the Indian boy did not move or make a sign, and Jeb realized that, of course, he did not understand English. After a few moments, however, the Indian's bow arm relaxed as he saw that Jeb was completely unarmed. Slowly and carefully, he replaced the arrow in his quiver and rearranged the position of his bow. He did all this without smiling or taking his eyes from Jeb. Little frown lines of worry creased his forehead, and Jeb felt that the Indian was trying to make up his mind what to do.

Then, still without any sign of friendship, the Indian lad pointed his finger at Jeb and then at a path that led down the steep hillside towards the village. The Indian was taking him captive. He could not risk letting Jeb go free to tell about the island. Jeb's heart sank and he wondered how long he would be held prisoner by the Indians. For the next few moments, his thoughts see-sawed back and forth between fear for his own safety and fear for Mac, alone and crippled on the beach. He knew he had to make the boy understand about Mac.

He motioned frantically towards the path that led to the beach, then to himself, and then back towards the beach again. The Indian scowled again and anger crept back into his eyes. He reached back to his quiver and rolled the shaft of an arrow back and forth between his thumb and forefinger.

Jeb shook his head in despair. He barked like a dog, and again pointed toward the beach. Then with a small stick he drew the outline of a dog in the earth. Jeb was not sure that the Indian understood him. Perhaps curiosity just got the better of him. But he did motion for Jeb to go down the path towards the beach. He also left no doubt in Jeb's mind that he would be following behind, guarding him closely all the way.

The walk back along the path was a strange, quiet one. Jeb walked quickly along, and the tall Indian followed noiselessly behind him. At last the path opening, leading to the white, sandy beach, was just ahead. Jeb hurried out onto the sand. Then, beckoning to the boy, he led him to the spot where Mac lay, still sick and weak. The dog's tail thumped eagerly on seeing Jeb, but a growl started low in his throat when he saw the stranger. Jeb spoke quickly to Mac. "Easy boy, quiet. It's all right."

The Indian bent over the dog lying on the soft

moss, and Jeb showed him the dog's wounded leg. Since the Indian boy seemed interested, Jeb sketched a crude picture of a boar in the sand with his finger, and once more pointed to Mac's bad leg. Then, he pointed to his boat to show how he had reached the island. After that, he drew a picture of a house and slowly shook his head back and forth to show that he did not know where his home was. The boy's expression did not change, and Jeb wondered if he was getting across to the Indian that he was lost, with a sick dog, and that he needed help.

For a moment, the two boys looked directly at one another. Still, Jeb could read nothing from the Indian's face. It was inscrutable, showing neither hostility nor friendship. Then, the Indian motioned for Jeb to follow him to the water to help him gather the coarse leaves of marsh grass growing near the shoreline. Perplexed, Jeb wondered what it was all about. But in a little while he understood, for the deft, agile fingers of the Indian boy soon wove a tight mat of the grass. Then, pointing to Mac, the Indian indicated that they should put the dog on the mat. Jeb talked quietly to the dog while he gently lifted him to the green litter. Jeb and the Indian each took an end. With Jeb in front, so the Indian could keep him in view, they went back along the trail through the woods. Jeb felt that so far he was fortunate, for he had not been harmed, and Mac was with him.

But above all else, there was the wonder of finding the answer to the mystery of the swamp, and knowing that it was not something evil that had caused him so much concern and fear. How simple, how wonderful, to have found an Indian village deep in the heart of the great Okefenokee. All the Indians had not been driven from the swamp years ago, as the early settlers had thought. Some had gone deeper into the Okefenokee and had lived on this beautiful island for many, many years. Probably afraid first of invading, warlike tribes, and later of the settlers, they had stayed hidden where others feared to go. The swamp was really two worlds—that of the swamp folk, like Jeb and his family, and that of the Indians.

But there had to be a boundary to the two worlds, and Jeb had discovered that boundary the day that he heard the sounds from the Indian hunting arrows.

Trudging on, Jeb considered his position. Although he did not relish the thought of being held captive, he did feel a little relieved that Mac and he would have shelter for a time. Perhaps there would be some good food for Mac, for both of them. Maybe he could make the Indians understand that he did not want to reveal their secret, and they would let him go. Perhaps they would even guide him out of this part of the swamp and back into his own world again. The thought cheered him, and with lifted spirits, he went on through the woods.

Abruptly, the two boys were again on the open hill overlooking the quiet village. They paused briefly before descending the hill. It was steep and difficult going with the sick dog, but soon they were approaching the village. With a nod of his head, the Indian motioned him towards the house from which, earlier, Jeb had seen smoke coming. Near the house the Indian beckoned for Jeb to stop. Carefully, they laid the tired dog down. Then, once more, the Indian boy held up his hand to indicate that Jeb should go no farther. He was to wait at that spot. The boy himself disappeared into a nearby home.

While Jeb stood outside, he looked around. Up close the houses did not appear to be as new and as well kept as they had from a distance. In fact the one the boy had disappeared into seemed to be in much better condition than the others. From that house Jeb now heard a long flow of Indian words, and then came the high, startled voice of a woman. Again, more words, and soon an old Indian woman appeared at the door. Her face was wrinkled and her body was bent. She seemed almost ancient, and Jeb decided that she must be the boy's grandmother. He could not help but notice the resentment that smoldered in her eyes as she looked at him, and he felt himself flushing. Again he realized that here on this island, he was the intruder.

As the old woman spoke, Jeb heard the name of the Indian boy for the first time. The grandmother called him "Tilo"; Jeb knew it was the boy's name by his response. Silently, Jeb formed the name with his lips so he would remember it.

Tilo spoke soothingly to his grandmother, who was highly agitated at seeing a stranger. Strange and hostile people had driven her ancestors here years ago, and she knew well the story of her tribe's search for a safe place to live. Gradually, the grandmother responded to the calming effect of Tilo's words. But she watched Jeb constantly with her intent, dark eyes.

As he watched them, Jeb realized how tall the boy and his grandmother were. Though Tilo seemed to be about the same age as Jeb, he was already half a head taller; and the old grandmother, even though bent with age, was close to six feet tall.

But what bothered Jeb was the stillness and quiet about the rest of the village. Why was smoke coming out of only one dwelling? Why was no one else around? The rest of the villagers should have been gathering by now to talk and exclaim over a stranger. But nothing happened. Nobody came. He wondered if the men were on a hunting party. But if that were the case, where were the women and children? Jeb turned again toward the cluster of silent homes, but nothing stirred and no one appeared. It seemed almost like a deserted village. Still he was thankful for the quiet; perhaps he would not have to face the tribal elders until the next day.

For the time being, Jeb had no more time for speculation. The grandmother was beckoning Jeb to enter their home. Jeb bowed and smiled a greeting, but he saw only the same guarded watchfulness on her face. Then, silently, the two boys carried

Mac into the house and laid him before the fire, where he sighed with comfort and closed his eyes.

Tilo's grandmother went over to the fire and removed a large pot. A wonderful aroma filled the air, and Jeb felt that he could hardly wait to eat. But as the grandmother scooped out bowls of hot stew, he was not so hungry that he could not see the dancing figures of the painted deer that leapt around the inner lip of the great pot. With a flash of memory, he could see the same kind of lighthearted deer on an ancient pottery bowl in his own cabin. It was the bowl that Dr. Bowen had taken from the mound on Pine Island.

Jeb had never tasted anything as good as the old grandmother's stew. He ate until he felt he could hold no more, and the grandmother appeared pleased that Jeb liked her food. Tilo could now approach Mac easily, for the dog had accepted him as a friend during the long walk from the beach to the village. Mac ate some of the good stew that Tilo took to him, and Jeb knew that his dog was recovering fast, for a sick dog will not eat. Even the wounded leg looked better, and when Mac shifted his position, it was plain to see that he was gaining back his strength.

Jeb looked around the Indian house. It was simple and clean and neat. There were woven mats on the floor, and piles of animal furs that were used for sleeping. Dominating the house was the fire in the center of the room, a symbol of the home.

Although Jeb had slept soundly in the punt, he was ready to sleep again when Tilo and his grandmother fixed soft furs on a mat for his bed and for their own. Jeb was not so sleepy, however, that he did not notice Tilo lay his bow and quiver close to his pallet before lying down himself. Evidently he intended to be on guard even in the night. The grandmother, too, seemed watchful and even restless. Jeb felt sorry for them. He tried to think of some way to reassure them, but sleep came on before his mind had found a solution.

When Jeb opened his eyes the next morning, Tilo and the grandmother were seated on mats near the fire and, at a sign from Tilo, he quickly took his place on an empty mat beside them.

For a few seconds, Tilo and his grandmother sat motionless, staring into the fire in an attitude of reverence. Then, Tilo reached into a large bowl on the floor beside him and took out a pinch of corn meal, which he threw into the fire. The powdery meal exploded in a shower of flames and, for several seconds longer, no one moved or spoke.

After this ritual, Tilo handed the jar of corn meal to his grandmother who tilted it over a pot of bubbling water, emptying the last bit into its steamy depths. Jeb noticed the quick look of concern that passed between Tilo and the grandmother and he wondered if it had anything to do with the empty corn jar. But he brushed the thought aside, for surely the fields behind the houses produced enough corn to fill many such jars. With only a few words passing between the two Indians and a few nods and grunts in Jeb's direction, each in turn drank the thin, hot corn gruel from a long, wooden ladle. Then, after giving Mac a portion, they finished the rest of the stew, which had been left simmering over the fire.

With the meal over, Jeb guessed that he would soon have to face the village council. When Tilo approached, motioning for Jeb to precede him out of the house, a little shiver of apprehension went through him, for he thought that now Tilo was going to take him to the elder tribesmen. His fate would be determined shortly; perhaps in the tremendous, round building behind the town. For Jeb suspected from its large size and prominent position that it was the village council house.

Out in the open, Jeb was once more aware of the awful stillness that hovered over the town. He had thought yesterday that the village looked deserted, but he had assumed the men and their families would soon return.

Yet, this morning, it was still empty. Still silent. No men or women worked about their daily tasks. No children laughed or played. A tiny suspicion, far back in his mind, was fanned to life. It grew and grew until, with a great burst of insight, Jeb understood. He knew! Except for Tilo and his grandmother, this was a deserted village. There were no children, or women, or hunters. There were no high priests or elders in the council house. There was only Tilo and the old grandmother, and Jeb—just the three of them—alone, in a village that was almost dead.

A great wave of sympathy rose in Jeb for Tilo and the old grandmother. And mixed with his sympathy was a puzzling question about the fate of the other villagers. What had happened to the rest of the tribe?

Jeb was awakened from his musings by Tilo who was pointing with a couple of long, slender sticks toward a region beyond the houses.

In a short time, the Indian boy had directed Jeb to the garden plot with its rows of corn and squash. Once there, he handed Jeb one of the slender sticks with its pointed, fire-hardened tip and, using the other, showed him how to cultivate around the plants.

From the hilltop, the garden had appeared to thrive; but now that he was in it, Jeb saw that it was a sorry looking sight. Large gaps existed between the corn plants where browsing deer had flattened the stalks; their hoofprints were everywhere in the soft soil. Even the standing stalks had some ruined ears, where deer, or crows, or raccoons had torn away the husks to get to the sweet kernels.

Again, Jeb felt compassion for Tilo, who was the sole provider for himself and his grandmother. In fact, Tilo was doing both a man's and a woman's work; for gardening was a chore for Indian women. He saw that because Tilo had no one to help protect the cornfield, at night or while he was away hunting and fishing, the corn yield was pitifully low. Now Jeb understood the look that had passed between Tilo and his grandmother over the empty corn jar. For while Tilo could always get fish and meat in the swamp, they also needed corn.

Jeb longed to help Tilo, and he thought of the large sack of corn meal in his own cabin. But his cabin was far, far away in a different world. Still, Jeb made a solemn promise that if he ever could, he would give Tilo and his grandmother some meal and a good supply of seed corn, too. For the present, however, Jeb knew that the most important thing he had to do was to find some way to break down Tilo's fear and suspicion. But try as he would, Jeb could think of no way to make Tilo trust him. For now, he would have to be content with showing Tilo that he was willing to help him with the work.

When the corn plants were cultivated, Jeb rested on his pointed stick, and wondered why Tilo was searching the ground so carefully. He appeared to be studying the deer-prints that crisscrossed the field.

At one place, Tilo seemed to have found what he was searching for, and, seeing the look of bafflement on Jeb's face, he pointed to several sets of tracks that were much clearer and sharper than the rest. Then, imitating an archer with a drawn bow, and pointing first to himself and then to Jeb, he indicated that they were going hunting. Jeb's eyes sparkled with excitement as he thought of going on a hunt. But he lost some of his zeal, when he realized that he would probably just tag along so that he could be under Tilo's watchful eye. For he knew with a certainty that Tilo would not trust him with a hunting bow.

Back at the house, Tilo helped Jeb to make ready for the hunt. He gave him a pair of his own moccasins to put on, and showed him how to drape a deerskin over his head and shoulders for a disguise. Then, covering his own body with a large deerskin, he picked up his bow and quiver and stopped to say a few words to his grandmother. The old lady nodded her head and smiled, although Jeb sensed a certain tension between them. The food was evidently needed. It would be good when the empty stewpot was full again.

Tilo and Jeb took up the hunt at the edge of the garden plot, and followed the deer-prints back through the woods. At one point, farther along the trail, Tilo paced off a length of about thirty feet and showed Jeb that he was to stay that distance behind him.

From then on, Jeb watched Tilo the hunter with great admiration. The Indian darted from tree to bush to tree, nimbly and with great skill. Never once did his moccasined feet make an undue sound upon the forest floor. Not once did a snapping twig or rustling leaf reveal his presence. But while all this was going on, Tilo kept his keen eyes on Jeb, who followed slowly and carefully behind at a distance. With great cunning and skill, Tilo made short advances, then took cover behind a tree or dense thicket, while working his way towards the small clearing up ahead. Nearer and nearer, he leaped stealthily towards the open spot, until, at the very edge of the clearing, he dropped quickly to the ground and hid from sight.

For a second or two, Jeb could not make him out among the forest shadows. Then he saw him, crouched and partly hidden by an old fallen tree. With fascination, he watched Tilo fit an arrow to his bow, draw back the bowstring, and carefully take aim at a deer within the clearing.

It was while he watched with admiration the Indian boy's skill, that Jeb's gaze also caught something else. A movement in the tree. Without hesitation Jeb screamed. And then he screamed again. He heard a crashing noise as the deer ran away, and he heard Tilo as he twisted around sharply to face Jeb. But he had only a fragmentary glimpse of Tilo's still-drawn bow, and his fierce, angry countenance as he glared at Jeb who had spoiled the hunt. For Jeb was frantically pointing and motioning towards a tree, just a short distance from where Tilo crouched.

Instinctively, Tilo twirled about and saw the wildcat Jeb had seen poised to spring. In the time before Tilo shot his arrow, Jeb thought incongruously of the miniature wildcat on Ma's kitchen shelf.

But this was not a carved, wooden animal. It was three feet of yellow fury, tensed to leap. The small ear tufts and snarling, open mouth gave it a sinister look. The wildcat's piercing scream and Tilo's snapping bowstring blended together, and cat and arrow met in mid-air.

For a brief interval, the woods were silent. Then both Jeb and Tilo rushed up to the body of the wild-cat, exclaiming together in an excited babel of English and Indian words. Neither could understand what the other was saying, but they did understand the silent language in the other's eyes. Tilo, in his anxiety to watch Jeb, had forgotten some of his customary caution; and Jeb had been his eyes of watchfulness. Now, the time had come for reappraisal, and Jeb and the Indian looked at one another searchingly. Then, simultaneously, both smiled and clasped hands. They were no longer captor and captive, but two brothers. For Jeb's action, which had saved Tilo from the sharp claws of the wildcat, had created a strong bond between them. The period of strain was over.

The boys shared the burden of carrying the thirty-pound wildcat back to the house, where Tilo quickly skinned the animal and left the meat for the old grandmother to cut up and cure.

Afer giving Mac a little attention, the two set off, in a spirit of comradeship, to set some snares and tend the traps. First, they set a few snares in runways between some nearby rocks. Then, on the way to set more snares, they stopped a while to shoot some arrows at a stump. When Jeb's turn came, he stretched the sinew bowstring far, far back; then thrilled to hear the arrow singing through the air.

Once more, they hurried on to check some larger traps along the forest's edge. Passing by a grove of oaks, Jeb stopped abruptly, for he was gazing at the village burial ground. Almost hidden, among the grove of trees, were over forty graves—mound after mound—just like those that were scattered on the larger islands throughout the swamp.

Tilo searched Jeb's face seriously; then beckoned for Jeb to listen to him. He seemed to feel the need to talk. He used his Indian language and also drew pictures in the earth for Jeb to see. And Jeb could understand a story that was sad and almost over. Tilo drew stick pictures of men and women who were strong and straight. Then, one by one, the pictures Tilo drew showed the strong men and women sick and lying down, until the only people left standing in the picture were a little boy and one tall Indian woman. Tilo indicated that the little boy was himself and the woman was his grandmother, now old and bent. In just a few minutes he had traced, in the earth, the sad history of his tribe. The pain in the Indian boy's eyes bit sharply into Jeb. In recalling the tragic tale, a bone-deep sorrow had stirred anew in the Indian boy. A sorrow that ached like an old wound.

Now there was no doubt at all in Jeb's mind. Tilo and his grandmother were the last members of the mound-builders tribe, and this island was their home. When his grandmother joined her Indian ancestors, Tilo would be all alone. Then, he would be the last of his race. The last of the giant race of beautiful Indians, so long only a legend.

Jeb and Tilo were silent as they made their way to a little woodland stream, for both were thinking of the sad story Tilo had just told. A short distance down stream, Tilo found what he was looking for. He stood there, gazing down at a fish-trap in the water, shaking his head because it was empty. Jeb peered into the water, too, and saw, immediately, that it was exactly like the one he had found on the divided island. Excitedly, with Tilo kneeling beside him, Jeb drew the outline of that island on the ground. Then he traced a little line down the middle, showing where the creek separated it into two halves. When he pointed to the place where he had found the fish-trap, Tilo nodded his head in surprise, and seemed almost as excited as Jeb. For the first time, the two boys had discovered a subject that they both knew something about.

Suddenly, Tilo jumped up and pointed back towards the village. Jeb wondered what there could possibly be in the village that had anything to do with the fish-trap or the island where he had seen it. Still puzzled, he followed after Tilo's racing feet until, flushed and out of breath, they reached the terraced incline leading upward to the great council house. Would Tilo take him there? Never had Jeb wanted to see the inside of a building so much. He was afraid that at any moment, Tilo might tell him to come no farther, but he continued to wave him on and together they climbed the steps cut into the four steep terraces. Once on the flat, open area, Tilo hurried Jeb past the courtyard with its wooden obelisk.

Finally, the two boys stood before the door of the great rotunda. Then Tilo pushed it open and they entered. Jeb never knew how long they paused there in the doorway. Tilo stood quietly, gazing straight ahead as though caught up in memories of the past, while Jeb looked eagerly all about the large, circular hall, memorizing every detail.

The thick wall of the great, hollow mound rose ten feet above the level of the floor, and had carved in it two tiers of circular benches on which the men and boys had sat during councils and religious ceremonies. Still covering the earthen benches were mats, and skins, and furs.

Embedded in the top of the mound wall was another wall of wooden posts, which raised the height of the structure another ten feet. This wooden wall had been plastered with a pale, yellowish clay and all around it, painted in bright, clear colors, were the dancing figures of Indian men who wore headdresses of animals such as bears, foxes, wildcats, and deer.

Looking still higher, Jeb saw that notches, cut

into the top of the wooden wall of posts, held the beams and rafters of the tremendous arched roof, which was covered with thick shingles of cypress bark. Supporting the center of the arched ceiling was a gigantic wooden pillar, which had been carved, from floor to ceiling, with frightening figures of writhing snakes. A little way out from the base of the great pillar was the fire-pit, still filled with cold, white ashes, and on the earthen floor, around both pit and pole, was a path worn by the feet of countless dancers.

Jeb's eyes were drawn next to a wooden altar that was set back a few feet from the pillar and fire-pit. It was plastered over with the same yellow clay as the walls, and was decorated with symbolic lines and patterns of sun and moon, dark streaks of thunder, and jagged lightning bolts.

On the top of the altar, in the exact center, sat a large fan of white eagle feathers, and arranged around it were numerous shallow bowls of colored powders and small skin pouches of charms and fetishes. At one end of the altar were two large conch shells, coated inside with the remains of a black liquid.

Standing there in the quiet room, Jeb's imagination began to play strange tricks. He saw the empty benches crowded with men and boys. Their low, muted voices filled the room. Then, almost in unison, they strained forward, while a great hush fell over the chamber. Anxiously, they watched the high priest light the cane heaped in the fire-pit. The cold, white ashes glowed with ruby lights; then scarlet flames leaped higher and higher, twisting upward like the carved serpents on the center pole. The priests began their chants. Over and over again, they repeated their monotonous chants, while the musicians started a slow beat with drums and rattles. Then dancers, with painted bodies and the headdress of animals, picked up the rhythm, stamping it out on the hard, dirt floor. Round and round the great center pole and fire-pit they danced. Faster and faster, while their frenzied bodies twisted and turned. Faster and faster, till the sweat gleamed on their naked backs. And always there was the chanting. On and on the priests chanted, imploring the dark and evil spirits of sickness to leave the village. Conch shells, filled with a thick, black liquid, were passed around and everyone drank to purify himself. Then, while the hypnotic flames leaped higher and higher, the priests threw in charms wrapped in small skin bags. Mysterious charms of powdered herbs and roots, and strange fetishes of feathers, horns, and bones that the flames quickly devoured.

Then Tilo touched Jeb's arm, and, in a flash, he was jerked out of his dream. Once more the room was quiet, the benches empty. Only cold, white ashes graced the fire-pit. And the only dancers were the painted ones, caught motionless on the high, plastered walls.

The dream was over, but for a little while, Jeb

had seen the chamber as it used to be, before a terrible sickness wiped out an entire village.

Tilo touched his arm again, and he remembered that the Indian boy had brought him here for a purpose. There was something in the great chamber that Tilo wanted to show him, something that had to do with the island where he had found the fish-trap.

Tilo led Jeb to the wall across from the doorway, where a shaft of light from the outside illuminated a large square of deer-skin that was set in a shallow, recessed area of the wall. The skin had been stretched taut over a wooden frame to form a canvas, and painted on the canvas, in beautiful, bright pigments, was a picture of the mound-builders' world.

Fascinated, Jeb saw that it was a square world, bounded on all four sides by straight stick figures of the four wind spirits—north, south, east, and west. In the top center was the tribal island, adorned by the artist with small houses, miniature stick people, and the great council house.

Then there were four other islands that were the hunting grounds. By hunting on first one and then the other, they had probably kept the game from getting too frightened or too scarce in any one vicinity. Now Jeb saw them as stepping stones that stretched across the canvas from bottom to top. Giant stepping stones, all but one of which he had touched briefly before reaching the mound-builders' beautiful island.

He was aware that Tilo was eagerly pointing to one of them, and Jeb saw that it was undoubtedly the island where he had found the fish-trap, for the painter showed it cut into two long halves by a blue stream of water. Jeb studied the other three islands one by one. Two meant nothing to him, for they lacked distinguishing features. Although one of them must have been the boar island. But the fourth one, near the border of the south wind, made him flush with excitement. It was a small, perfect oval island, set in a ring of bright green water. It was the Emerald Isle.

Jeb continued to gaze at the picture while he thought about the strange twist of events. Tilo had brough him here to show that he knew where he had found the fish-trap. But now, Jeb saw that the picture provided him with a key that would unlock the door of the swamp's wild heartland, so that he could return quickly to his own world again. For by using this picture, he could make Tilo understand that he wanted him to guide him to the Emerald Island; from there, he could easily find his way home.

Eagerly, Jeb scraped the forms of two canoes in the earthen floor and, with appropriate gestures, showed Tilo that he wanted him to lead him back to the border of their two worlds—back to the Emerald Island.

Jeb saw that Tilo understood him, but the Indian boy sadly shook his head back and forth. Again and again, he swept his hand over the picture and then placed a finger over his lips. By tribal law, the island had to remain secret. He could not let Jeb reveal the mound-builders' world.

Jeb leaned forward. His eyes were intent and he was trembling a little. He had to show Tilo that he could be trusted to keep the island a secret. Three times, he swept his own hand over the picture and then placed his finger over his lips, and then over his heart. Impulsively, he took a step toward Tilo and, reaching out, clasped the Indian boy's hand in his. Perhaps, Tilo remembered when they had clasped hands once before in friendship, after the incident with the wildcat, or maybe it was just that he recognized the sincerity in Jeb's face, for after a few seconds of deep thought, he smiled and shook his head up and down. He trusted Jeb to keep the tribal secret. He would take him to the Emerald Isle.

On the walk back, through the silent village, to the house, Jeb thought about his stay on the island. He had eaten and slept in a mound-builder's home, and had walked through a whole day of village life in their moccasins. Tomorrow, he would be gone from the island, but he knew he would never forget Tilo and his grandmother. Again, he vowed that he would meet Tilo again and give him a gift of corn. And far back in his mind, a little dream was taking shape. A dream that Tilo and he would meet often enough to learn to speak one another's language so they could really understand one another.

13

AFTER BREAKFAST THE NEXT MORNING, THE TWO
boys realized that the time was drawing near when
Jeb and Mac would have to leave. Tilo and Jeb had
been drawn together by a deep tie of friendship, and
neither boy wanted to part from the other.

Tilo picked up a large pottery jar and Jeb
walked with him to a shallow cave formed in one of
the steep walls, rising up from the valley floor.
There, Tilo filled the jar with water from a spring
that seeped down through the rocky hillside. It was
here, at this spot, that Tilo stopped and seriously put
his hand on Jeb's shoulder in a token of friendship,
and the two boys again renewed their solemn pledge

of secrecy. Silently, Tilo took from around his neck a small figure of an animal—a good luck charm that he had worn all his life to protect him from evil spirits. Jeb was deeply touched when Tilo fastened it around his neck.

Jeb wanted desperately to give Tilo something in return, but at first he could think of nothing. Then, he remembered his pocketknife, which he often carried so he could whittle in his spare time. Now, he pulled it from his pocket and showed Tilo how to open the three shiny blades. Tilo accepted the knife with delight.

Soon it was time for the long journey home. After good-byes were said to the old grandmother, Tilo and Jeb carried Mac on the litter back to the sandy beach and Jeb's boat.

Mac showed renewed strength and interest, and was able to half sit in his place in the bow of the punt. Jeb let his boat drift in the shallow water while he waited for Tilo to get his dugout canoe from its hidden mooring place farther down the beach.

The two boats sped gracefully through the strange, jungle-like beauty of the swamp forests, and then darted out into the open meadows and prairies like two carefree dragonflies. Each boy skillfully guided his narrow boat in and out of countless winding channels.

It had been a long journey to the beautiful island, deep in the heart of the great swamp, and it was a long journey back to the boundary of the swamp world that Jeb knew so well. Not until they reached the green tunnel, did Jeb recognize any of the landmarks of the unknown swampland. It had been a strange journey that seemed like weeks instead of the few days it had actually been.

At the green tunnel, Tilo drew his boat to one side and stopped.

He and Jeb looked solemnly at each other, and between them a message flashed, a wordless exchange that was both a farewell and a beginning. Then they repeated their symbolic gestures of friendship.

"Good-bye, Tilo. Thank you." Jeb didn't know whether Tilo would recognize his name when he spoke it or not, but Tilo's quick smile showed that he did.

Without looking back, Jeb shoved his little boat into the green tunnel of plants that led to Emerald Lake.

It had been a strange adventure and one that Jeb knew he would not be able to tell about. For while the old grandmother lived, Tilo would have to provide for her and protect her, living just as his people had lived for hundreds of years, until the year of the great sickness had caused most of them to die. And as far as Jeb was concerned, no one was going to know they were there.

But Jeb knew that he would go often to Emerald Island, and he believed that Tilo would go there, too. For their friendship was strong and they would want to see each other again.

It was curious how good all the familiar landmarks looked to Jeb on his journey back to his cabin. Many times the moss from the cypress trees brushed against his face, but it no longer felt cold or unfriendly. Even the many animal noises seemed to be welcoming him back. And Jeb knew that never again would an unknown sound cause him to panic in his swamp world.

Someday, Jeb realized, Tilo would be all alone, and no man can live alone forever. Jeb would be his only friend. And it would be he, Jeb, who could lead Tilo gently from an old way of life to a new one. Tilo was young and with Jeb for a friend, he would be able to bridge the gap between the two civilizations.

Someday, Tilo would come to live with Jeb as his brother, and together they would finish growing up in the great swamp, which they both loved so well.

With the sun low in the sky, Jeb skillfully turned his little boat into the place where the creek and the swamp merged and left the gentle waters of the Okefenokee for the swifter waters of the creek.

Now Jeb knew that home was not far away. Home with its rows of shining, white beehives, and its beloved, old log house, safe within the arms of the split-rail fence. Jeb would be happy to see them all.

But best of all, Ma and Pa would be waiting.

They would welcome him home from his long journey, glad that he had returned. But they would not question him about his stay in the swamp, for they knew that swamp men like to be away by themselves.

Suddenly, throwing back his head, Jeb's yodel pierced the still evening air and echoed far up the creek. Ma and Pa heard it and knew that he was coming home. Nod, his little pet alligator, heard it, too, and moved his small body imperceptibly on the dock. Waiting.